Eduqas
Physics

A level Year 2

Study and Revision Guide

Gareth Kelly

Nigel Wood

Iestyn Morris

Published in 2017 by Illuminate Publishing Limited, an imprint
of Hodder Education, an Hachette UK Company, Carmelite House,
50 Victoria Embankment, London EC4Y 0DZ

Orders: Please visit www.illuminatepublishing.com
or email sales@illuminatepublishing.com

British Library Cataloguing in Publication Data

A catalogue record for this book is available from the British Library
ISBN 978-1-908682-73-4

Printed by Severn, Gloucester

10.21

The publisher's policy is to use papers that are natural, renewable and recyclable
products made from wood grown in sustainable forests. The logging and manufacturing
processes are expected to conform to the environmental regulations of the country of
origin.

Every effort has been made to contact copyright holders of material reproduced in this
book. If notified, the publishers will be pleased to rectify any errors or omissions at the
earliest opportunity.

This material has been endorsed by Eduqas and offers high quality support for the
delivery of Eduqas qualifications. While this material has been through a Eduqas quality
assurance process, all responsibility for the content remains with the publisher.

WJEC examination questions are reproduced by permission from WJEC.

Editor: Geoff Tuttle
Cover and text design: Nigel Harriss
Text and layout: Kamae Design, Oxford

Acknowledgments

We are grateful to the team at Illuminate Publishing for their support and guidance
throughout this project. It has been a pleasure to work so closely with them.

Thanks also are due to Keith Jones for his many supportive suggestions about the text and
for an anonymous worker who meticulously checked our answers to questions.

Contents

How to use this book

As examiners and former examiners we have written this new study guide to help you be aware of what is required for – and structured the content to guide you towards – success in the year 2 part of the Eduqas A level examination in Physics.

There are three main sections to the book:

Knowledge and Understanding

This first section covers the key knowledge required for the examination.

You'll find notes on the contents of the examination components:

- Component 1: Newtonian physics
- Component 2: Electricity and the universe
- Component 3: Light, nuclei and options

which are not covered in the Year 1 of AS SRG including the practical and data-handling skills which you will need to develop.

In addition there are a number of features throughout this section that will give you additional help and advice as you develop your work:

Component introduction

The key sub-sections are listed with their page references and their corresponding exam questions. Each then has a short summary giving you an essential overview of the area of study, plus a revision checklist as you work through your revision process.

- **Key terms**: many of the terms in the Eduqas specification can be used as the basis of a question, so we have highlighted those terms and offered definitions.

- **Quickfire questions**: are designed to test your knowledge and understanding of the material.

- **Pointer and Grade Boost**: offer extra examination advice based on experience of what candidates need to do to attain the highest grades.

- **Extra questions**: feature at the end of each topic or area of study providing you with further practice at answering questions with a range of difficulty.

Practical and data-handling skills

This section deals with the practical and data skills you will need to demonstrate in both the non-examination assessment (NEA) and in the component papers. It builds upon the information given in the AS level Study and Revision Guide.

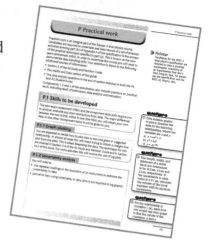

Exam Practice and Technique

The third section of the book covers the key skills for examination success and offers you examples based on suggested model answers to possible examination questions. First, you will be guided into an understanding of how the examination system works, an explanation of Assessment Objectives and how to interpret the wording of examination questions and what they mean in terms of exam answers.

A variety of structured practice questions, including a QER question, is provided, taken from across the year 2 part of the specification. Model answers are given. This is followed by a selection of examination and specimen questions with actual student responses. These offer a guide as to the standard that is required, and the commentary will explain why the responses gained the marks that they did.

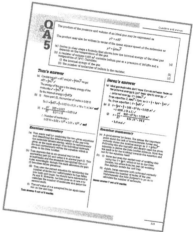

Most important of all, we advise you to take responsibility for your own learning and not rely on your teachers to give you notes or tell you how to gain the grades that you require. You should look for extra reading and additional notes to support your study in physics. It's a good idea to check the awarding body website – www.eduqas.co.uk – where you can find the full subject specification, specimen examination papers, mark schemes and in due course examiner reports on past years' exams.

Good luck with your revision!

Component 1 Knowledge and Understanding

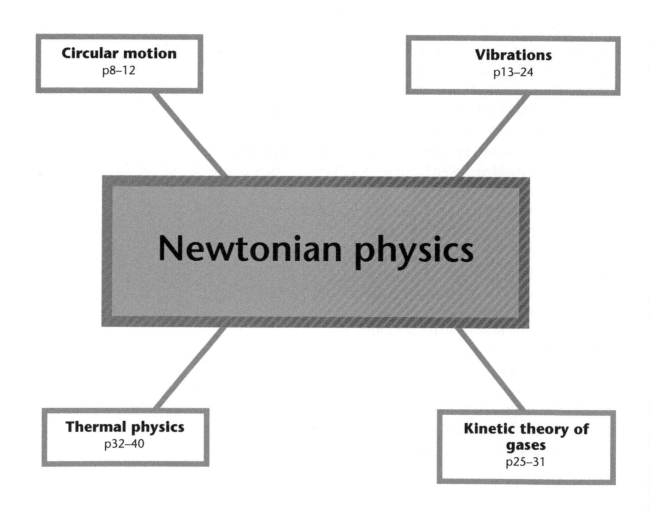

Circular motion
p8–12

Vibrations
p13–24

Newtonian physics

Thermal physics
p32–40

Kinetic theory of gases
p25–31

Circular motion

The kinematics and dynamics of objects moving in a circle; the concepts you will need to handle the motion of objects from fairground rides and slingshots to planets and galactic stars

p8–12

Vibrations

Simple harmonic motion (shm); its mathematical and graphical expression; its application to spring systems; the response of objects to periodic driving forces including resonance in structures such as bridges.

p13–24

Kinetic theory of gases

The behaviour of gases in terms of their molecules; the equation of state for ideal gases; the assumptions and predictions of the kinetic theory of gases; the molar gas constant, the Boltzmann and Avogadro constants; the kinetic energy of gas molecules.

p25–31

Thermal physics

Thermodynamic systems; the first law of thermodynamics; the internal energy of a gas; p-V graphs; internal energy changes in solids and liquids; heat flow caused by temperature differences; specific heat capacity.

p32–40

Basic notes   Good grasp   Fully revised

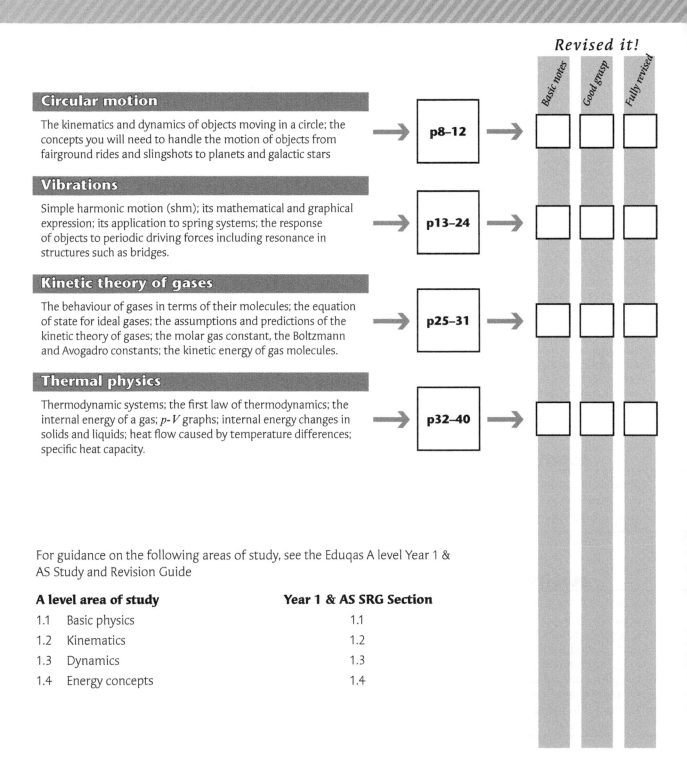

For guidance on the following areas of study, see the Eduqas A level Year 1 & AS Study and Revision Guide

A level area of study		Year 1 & AS SRG Section
1.1	Basic physics	1.1
1.2	Kinematics	1.2
1.3	Dynamics	1.3
1.4	Energy concepts	1.4

1.5 Circular motion

We deal with motion in a circle at constant speed, covering ideas of angle measurement and angular velocity, and how to apply $F = ma$ to uniform circular motion.

1.5.1 Defining an angle in radians

» Pointer

What's the point of radians? The degree is an *arbitrary* unit, because there's no *fundamental* reason for choosing 360 degrees in a revolution.

By contrast, the radian is a *natural* unit. Its use simplifies several angle-related formulae.

Imagine an arc of a circle, of any radius, r, centred on O, (where the 'arms', OA and OB, of the angle meet). See diagram. Then the angle θ is given, in radians, by...

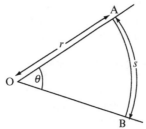

Fig. 1.5.1 θ in radians

$$\theta / \text{rad} = \frac{\text{arc length}}{\text{radius}}, \quad \text{that is} \quad \theta/\text{rad} = \frac{s}{r}$$

If $s = r$, then $\theta/\text{rad} = 1$, that is $\theta = 1$ radian.

Example

A ship in the Pacific Ocean cruises for 1200 km along the equator. What angle in radians at the centre of the Earth does the journey subtend? (Take the Earth's equatorial radius to be 6380 km.)

Answer

$$\theta = \frac{\text{arc length}}{\text{radius}} = \frac{1200 \text{ km}}{6380 \text{ km}} = 0.188 \text{ rad}$$

In practice you'll probably never have to *measure* an arc length in order to determine an angle in radians. It's important, though, to be able to convert between degrees and radians – either way round. Read on.

For a complete revolution, $s = 2\pi r$ so $\theta / \text{rad} = \frac{2\pi r}{r} = 2\pi$.

Thus, $2\pi \text{ rad} = 360°$, $\pi \text{ rad} = 180°$, $1 \text{ rad} = \frac{180°}{\pi} = 57.3°$ $1° = \frac{\pi}{180} \text{ rad}$

Example

What is 60° in radians?

Answer

$$60° = 60 \times 1° = 60 \times \frac{\pi}{180} \text{ rad} = \frac{\pi}{3} \text{ rad} = 1.05 \text{ rad.}$$

The unit 'rad' is often omitted when there's no ambiguity. For example, you may write $\theta = \frac{\pi}{3}$, meaning $\theta = \frac{\pi}{3} \text{ rad}$, but don't write $\theta = 1.05$. In some of the formulae below, involving angular velocity, 'rad' is omitted.

1.5.2 Angular velocity

The **angular velocity**, ω, of a point moving at a constant rate in a circular path with its centre at point O is defined as:

$$\omega = \frac{\text{angle swept out (about O)}}{\text{time taken}}, \text{ that is } \omega = \frac{\theta}{t}$$

UNIT: rad s⁻¹

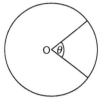

Fig. 1.5.2 Angle swept out

For one whole revolution, $\theta = 2\pi$, and $t = T$, the period of rotation, this being the time for one revolution (one cycle).

Thus, $\omega = \dfrac{2\pi}{T}$ that is $\omega = 2\pi \times \dfrac{1}{T}$ so $\omega = 2\pi f$,

in which f is the **frequency of rotation**: the number of revolutions, or cycles, per unit time.

Example

Calculate the angular velocity of the seconds hand of a clock.

Answer

$$\omega = \frac{2\pi}{T} = \frac{2\pi}{60\,\text{s}} = 0.105 \text{ rad s}^{-1}$$

Key Terms

Angular velocity ω

$$\omega = \frac{\text{angle swept out}}{\text{time taken}} = \frac{\theta}{t}$$

UNIT: rad s⁻¹

The **period of rotation**, T, is the time for one revolution.
UNIT: s

The **frequency**, f, is the number of revolutions per unit time.

UNIT: s⁻¹ = hertz (Hz)

$$T = \frac{1}{f} = \frac{2\pi}{\omega} ; f = \frac{1}{T} = \frac{\omega}{2\pi}$$

Scalar or vector?

In advanced work, ω is treated as a sort of vector (pointing along the rotation axis), allowing the handling of rotations about axes at various angles. At A-level we deal only with one axis direction in any problem, so we can treat ω as a scalar.

Speed and angular velocity

Suppose a body moves round a circular path at speed v. It traverses an arc of length $s = vt$ in a time t. But $s = r\theta$,

so $r\theta = vt$, that is $\dfrac{\theta}{t} = \dfrac{v}{r}$. So $\omega = \dfrac{v}{r}$ and $v = r\omega$.

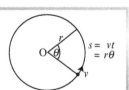

Fig. 1.5.3 Two expressions for s

quickpire

③ A wheel of diameter 0.48 m rotates at 3000 turns per minute. Determine (in SI units):
a) its rotation frequency
b) its angular velocity
c) the speed of a point on its circumference.

1.5.3 Centripetal acceleration

A body moving at constant speed around a circular path is accelerating, because its velocity is always changing (in direction).

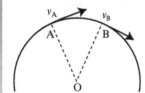

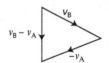

The change in velocity over the arc AB is found using the vector diagram, which is based on:

velocity at B – velocity at A
= velocity at B + (–velocity at A)

Fig. 1.5.4 Velocity change for body on a circular path

As the vector diagram suggests, the change in velocity, and hence the acceleration, is *centripetal*: always directed towards the circle centre.

For a body moving at speed v (and angular velocity ω) in a circle of radius r, the magnitude of the acceleration is

$$a = \frac{v^2}{r} \quad \text{or, equivalently,} \quad a = r\omega^2$$

Example

The *London Eye* is a giant wheel rotating at a constant angular velocity. Each revolution takes 30 minutes. Calculate the acceleration of a point on the wheel 60 m from its centre.

Answer

$$\omega = \frac{2\pi}{T} = \frac{2\pi}{30 \times 60\,\text{s}}; \quad a = r\omega^2 = 60\,\text{m} \times \left(\frac{2\pi}{30 \times 60\,\text{s}}\right)^2 = 7.3 \times 10^{-4}\,\text{m}\,\text{s}^{-2}$$

Centripetal force

A body can't move in a circle at constant speed without having a resultant force acting on it towards the centre of the circle, to give it the centripetal acceleration. Using Newton's second law,

$$F_{res} = ma \quad \text{so in this case} \quad F_{res} = m\frac{v^2}{r} \quad \text{and} \quad F_{res} = mr\omega^2$$

Example

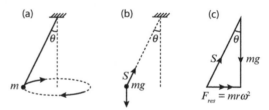

Fig. 1.5.5 'Conical pendulum' and the forces on its bob

> ### » Pointer
> Check that the units of $\frac{v^2}{r}$ are those of acceleration.

> ### » Pointer
> You should show, using $v = r\omega$, that $a = \frac{v^2}{r}$ and $a = r\omega^2$ really are equivalent.

quickfire
④ A train is travelling at 18.0 m s^{-1} on a curved section of track. The curve is an arc of a circle of radius 120 m. Calculate the train's acceleration.

quickfire
⑤ A carriage of the train in Quickfire 4 has a mass of 36000 kg. Find the centripetal force on it. What external 'object' exerts this force?

A ball, of mass $0.100\,\text{kg}$, is attached to a string and whirled in a *horizontal* circle of radius $0.30\,\text{m}$ at a rate of 1.1 revolutions per second. The string sweeps out a cone as in Fig. 1.5.5 (a). Calculate the angle of the string to the vertical and the tension, S.

Answer

The resultant force on the bob provides its centripetal acceleration,

So $\quad F_{\text{res}} = mr\omega^2 = 0.100\,\text{kg} \times 0.30\,\text{m} \times (2\pi \times 1.1\,\text{s}^{-1})^2 = 1.43\,\text{N}$.

This must be the resultant of the pull, S, of the string and the pull of gravity, mg, on the bob. See Fig. 1.5.5 (b) and (c). From (c) we deduce that

$$S = \sqrt{F^2_{\text{res}} + (mg)^2} = \sqrt{1.43^2 + 0.981^2}\,\text{N} = 1.73\,\text{N}$$

and $\quad \theta = \tan^{-1}\left(\dfrac{mr\omega^2}{mg}\right) = \tan^{-1}\left(\dfrac{1.43}{0.981}\right) = 56°$

quicKfire

⑥ In a 'conical pendulum' set-up (see example) the string is $0.60\,\text{m}$ long and is always at $30°$ to the vertical. The bob's mass is $0.100\,\text{kg}$. Calculate:

a) the tension in the string (consider vertical force components)

b) the radius of the bob's circular path

c) the bob's speed.

1. Fig. 1.5.6 shows a simple 'chain drive' on a bicycle.

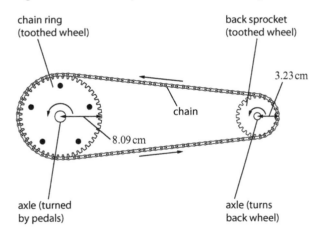

chain ring (toothed wheel)

back sprocket (toothed wheel)

3.23 cm

chain

8.09 cm

axle (turned by pedals)

axle (turns back wheel)

Fig. 1.5.6 Bicycle chain drive

A cyclist pedals so as to turn the chain ring at a rate of 1.50 revolutions per second. Calculate:

(a) the angular velocity of the chain ring

(b) the speed at which the links in the chain are made to move (relative to the bicycle)

(c) the angular velocity of the back sprocket (and back wheel)

(d) the speed at which the bicycle is travelling, if the diameter of the back wheel (to the tread of the tyre) is $0.67\,\text{m}$.

2. A washing machine has a drum (a metal cylinder open at one end and closed at the other) of diameter 0.51 m, mounted with its axis horizontal. On the spin-dry phase the cylinder rotates at 1500 revolutions per **minute**.

    (a) Calculate the drum's angular velocity (in SI units).

    (b) Calculate the centripetal force on a wet sock of mass 0.085 kg clinging to the inside wall of the drum.

    (c) *Hence* explain why gravity has almost no effect on the action of the dryer.

    (d) There are small holes in the curved surface of the drum. Explain why water droplets from the socks escape through these holes rather than staying in the sock.

3. A pendulum consists of a bob of mass m suspended from a fixed point by a string of length l. The string is held taut and horizontal, then the bob is released (Fig. 1.5.7).

Fig. 1.5.7 Pendulum

(a) Use the *principle of conservation of energy* to show that the bob's speed at its lowest point is given by
$$v = \sqrt{2gl}$$

(b) Hence express the centripetal force on the bob at its lowest point in terms of m and g.

(c) Calculate the tension in the string, for the bob at its lowest point, remembering that the centripetal force is the resultant of *two* forces on the bob.

1.6 Vibrations

When an object is moving to and fro regularly about a fixed point, we say that it is *vibrating* or *oscillating*. We'll deal mainly with two well-known examples: an object on a spring, and a pendulum. Both these systems can oscillate naturally with so-called **simple harmonic motion** (shm).

1.6.1 The definition of simple harmonic motion

It helps to have a particular system in mind. We'll consider the block on a spring, shown in Fig. 1.6.1. When the block, m, is displaced from its equilibrium position and released, it oscillates back and forth.

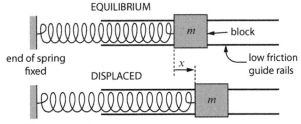

Fig. 1.6.1 Block oscillating on spring

When m's displacement is x, the spring's extension is x, and, according to Hooke's law, the spring exerts a force proportional to x on the block, given by:

$$F = -kx.$$

- The minus sign is because the force is in the opposite direction to x.
- k is the **spring constant**: see **Key terms**.

Assuming resistive forces are negligible, the block's acceleration is:

$$a = \frac{F}{m} \quad \text{so} \quad a = -\frac{k}{m}x \quad \text{which is often written as} \quad a = -\omega^2 x.$$

in which ω is a constant given by $\omega = \sqrt{\dfrac{k}{m}}$.

Oscillations obeying $a = -\omega^2 x$, that is $a = -$ constant $\times x$, are called *simple harmonic motion* (shm). This definition is put into words in **Key terms**.

The relationship is shown as a graph in Fig. 1.6.2. A is the **amplitude**: the maximum value of the displacement.

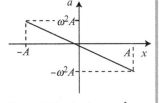

Fig. 1.6.2 Graph of $a = -\omega^2 x$

Key Terms

Simple harmonic motion occurs when an object moves so that its acceleration is always directed towards a fixed point and is proportional to its displacement from that point.

Mathematically:
$a = -\omega^2 x$

The **spring constant**, k is the force exerted per unit extension by a spring.

UNIT: N m^{-1}

» Pointer

When the block in Fig. 1.6.1 is to the left of the equilibrium position, the spring is compressed. As long as the turns of the spring don't touch, k will have the same value for compression as extension.

quickfire

① Calculate the gradient of a graph of a against x (see Fig. 1.6.2) for a body of mass 0.15 kg oscillating on a spring for which $k = 5.4$ N m^{-1}.

quickfire

② Show that the units of ω are s^{-1}. What quantity associated with oscillations has the same unit?

Key Terms

The **amplitude** of an oscillation is the maximum value of the displacement.

The **periodic time** (or period), T, is the time for one cycle.

The **frequency**, f, is the number of cycles per unit time.

UNIT: s^{-1} = hertz (Hz).

The **phase**, $(\omega t + \varepsilon)$, is the angle that tells us the point reached in the oscillation cycle.

Vertical oscillations of an object hanging from spring

The equation $a = -\omega^2 x$ in which $\omega^2 = \dfrac{k}{m}$ also applies to the vertical oscillations of m in Fig. 1.6.3. This is far easier to set up than the arrangement in Fig. 1.6.1, but note that x, the displacement from equilibrium, is not now the spring extension. Why not?

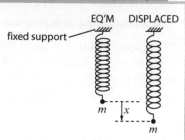

Fig. 1.6.3 Vertical oscillations

1.6.2 Variation of displacement with time for a body in shm

As we'll confirm later, a body will have acceleration given by $a = -\omega^2 x$ if (and only if) its displacement varies with time according to

$$x = A \cos (\omega t + \varepsilon) \qquad (\omega, A \text{ and } \varepsilon \text{ are constants.})$$

This motion is closely related to that of a point, **P**, moving with angular velocity ω round and round a circle of radius A. In fact, x is the horizontal *component* of P's displacement from the circle centre (Fig. 1.6.4), and ωt and ε are the angles shown.

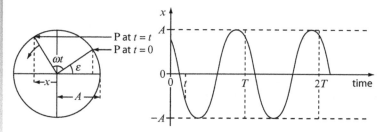

Fig. 1.6.4 $x = A \cos (\omega t + \varepsilon)$ as circular motion component

A is the **amplitude** of the shm: the maximum displacement. It is a constant determined by how the body is set in motion, for example by how far we displace it from its equilibrium position, before letting go.

The **periodic time**, T, the time for one cycle of oscillation, is the time for one revolution of point P – see Fig. 1.6.4. So, from Section 1.5.2,

$$\omega = \frac{2\pi}{T} \quad \text{that is} \quad T = \frac{2\pi}{\omega}.$$

Equivalently, as **frequency**, $f = \dfrac{1}{T}$ we have $f = \dfrac{\omega}{2\pi}$ that is $\omega = 2\pi f$.

- For a mass–spring system, since $\omega = \sqrt{\dfrac{k}{m}}$, we have $T = 2\pi \sqrt{\dfrac{m}{k}}$

- For a **simple pendulum** (small bob hanging by a light thread from a fixed support), swinging no further than about 20° from the vertical, the bob's acceleration is related to its displacement x along the arc by $a = -\omega^2 x$

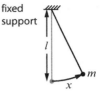

fixed support

- in which $\omega = \sqrt{\dfrac{g}{l}}$, so $T = 2\pi\sqrt{\dfrac{l}{g}}$.

$(\omega t + \varepsilon)$ is an angle called the **phase**. It tells us the stage reached in the oscillation cycle at time t.

ε is an angle called the **phase constant.** We can set its value so that time zero is at any point we choose in the body's cycle...

If we want $t = 0$ to be when $x = A$, we choose $\varepsilon = 0$, giving

$x = A\cos(\omega t)$. (Fig. 1.6.5 (a))

If we want $t = 0$ to be when $x = 0$ with x about to be positive we choose

$\varepsilon = -\dfrac{\pi}{2}$. Because $\cos\left(\omega t - \dfrac{\pi}{2}\right) = \sin(\omega t)$, we have

$x = A\sin(\omega t)$. (Fig. 1.6.5 (b))

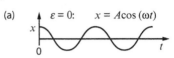

(a) $\varepsilon = 0$: $x = A\cos(\omega t)$

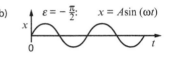

(b) $\varepsilon = -\dfrac{\pi}{2}$: $x = A\sin(\omega t)$

Fig. 1.6.5 Useful cases of $x = A\cos(\omega t + \varepsilon)$

Example 1

A metal sphere of mass $0.16\,\text{kg}$ hangs from a spring of spring constant $40\,\text{N m}^{-1}$. The sphere is released from rest with a displacement of $+0.030\,\text{m}$. Calculate the period of its oscillations and its displacement $0.55\,\text{s}$ after release.

Answer

$\omega = \sqrt{\dfrac{k}{m}} = \sqrt{\dfrac{40\,\text{N m}^{-1}}{0.16\,\text{kg}}} = 15.8\,[\text{rad}]\,\text{s}^{-1}$ So $T = \dfrac{2\pi}{\omega} = \dfrac{2\pi}{15.8\,\text{s}^{-1}} = 0.40\,\text{s}$.

$A = 0.030\,\text{m}$, because the displacement will oscillate between $\pm0.30\,\text{m}$.

$\varepsilon = 0$ if we call the release time (from maximum displacement) $t = 0$.

So $x = A\cos(\omega t) = 0.030\,\text{m} \times \cos(15.8 \times 0.55\,\text{rad}) = -0.022\,\text{m}$.

- To evaluate $\cos(15.8 \times 0.55\,\text{rad})$ directly, your calculator needs to be in *radian* mode (not degree mode).
- Use a sketch-graph to check that the answer is sensible.

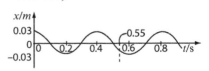

Fig. 1.6.6 Graph for Example 1

>> *Pointer*

The value of the constant, ω, is fixed by the system itself (for a mass-spring system by k and m).
But A and ε are determined by how the system is set in motion and timed.

>> *Pointer*

It's (mercifully) rare to need any other version of $x = A\cos(\omega t + \varepsilon)$ than $x = A\cos(\omega t)$ or $x = A\sin(\omega t)$.

▲ *Grade boost*

Don't be afraid to draw $x - t$ sketch-graphs. Sometimes they're essential.

quickɸɪʀe

③ A metal ball of mass $0.12\,\text{kg}$ attached to a fixed point by a spring performs 43 cycles of oscillation in a *minute*. Calculate the spring constant.

quickfire

④ ⑤ ⑥ ⑦

A metal ball hangs from a long thread, as a pendulum. The ball is pulled to one side from its rest position by 0.040 m in the +x direction and released. It performs shm of period 2.00 s.

quickfire

④ Calculate the pendulum length, l.

quickfire

⑤ Calculate the displacement, x, of the ball 1.20 s after release.

quickfire

⑥ Determine x at 1.50 s after release. [Easy by sketch-graph!]

quickfire

⑦ Calculate the times from release at which x is −0.020 m:
a) for the first time
b) for the next time.

Example 2

A piston in an engine is moving with shm, of amplitude 0.040 m, and frequency 35 Hz. At one point in its cycle its upward displacement from its mid position is 0.024 m, and is increasing.

(a) Find how long it takes to reach this point from its mid position.

(b) How much longer will it be before this displacement occurs again?

Answer

(a) Choose $\varepsilon = -\dfrac{\pi}{2}$, to make $t = 0$ when piston passes through its mid position, going up. (We'll call upwards the +x direction.)

Then $x = A\sin(\omega t)$, so $\sin(\omega t) = \dfrac{x}{A} = \dfrac{0.024 \text{ m}}{0.040 \text{ m}} = 0.600$

So $\omega t = \sin^{-1}(0.600) = 0.644$ rad (calculator in radian mode!)

So $t = \dfrac{0.644}{\omega} = \dfrac{0.644}{2\pi f} = \dfrac{0.644}{2\pi\,35\,\text{s}^{-1}} = 2.93$ ms. To 2 sf, $t = 2.9$ ms

A quick check: $\dfrac{T}{4} = \dfrac{1}{4} \times \dfrac{1}{f} = \dfrac{1}{4} \times \dfrac{1}{35\,\text{s}^{-1}} = 7.14$ ms;

2.93 ms < 7.14 ms; confirming that 2.93 s is during the first upstroke.

(b) The 'inverse sine' ($\sin^{-1}$) function on a calculator gives only the angle (with the given sine) in the range $-\frac{\pi}{2}$ to $+\frac{\pi}{2}$. To find any more times, *after* 2.9 ms, when $x = 0.024$ m, you need to draw a circle diagram or – as here – a sketch-graph (Fig. 1.6.7), making use of its symmetry.

The extra time elapsing is seen to be 8.42 ms (8.4 ms to 2 sf). Note the use of $\dfrac{T}{4} = 7.14$ ms.

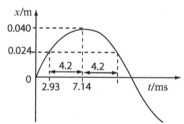

Fig. 1.6.7 Graph for Ex 2 (b)

1.6.3 Velocity of a body in shm

A displacement–time (x–t) graph is sketched (Fig. 1.6.8 (a)) for a body moving according to

$$x = A \cos(\omega t + \varepsilon).$$

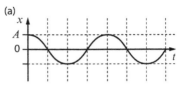
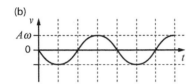

Fig. 1.6.8 x–t and v–t

For simplicity, the graph is drawn for the case $\varepsilon = 0$.

The gradient of the tangent to the x–t graph at any time gives the velocity at that time. Hence the shape of the velocity–time (v–t) graph (Fig. 1.6.8 (b)). In fact v varies with t according to

$$v = -A\omega \sin(\omega t + \varepsilon).$$

The values of sines (and cosines) range between -1 and $+1$. So the maximum velocity, v_{max} is $-A\omega \times (-1)$. That is:

$$v_{max} = A\omega.$$

Example

(a) A metal ball of mass $0.10\,\text{kg}$ hangs from a spring of spring constant $3.6\,\text{N}\,\text{m}^{-1}$. What amplitude of oscillation will give the ball a maximum speed of $0.21\,\text{m}\,\text{s}^{-1}$?

(b) At what displacement from the equilibrium position will the ball have half its maximum speed?

Answer

(a) $v_{max} = A\omega = A\sqrt{\dfrac{k}{m}}$ So $A = v_{max}\sqrt{\dfrac{m}{k}} = 0.21\sqrt{\dfrac{0.10}{3.6}}\,\text{m} = 0.035\,\text{m}$

(b) Taking ε as zero (for simplicity) and remembering that $v_{max} = A\omega$, we have

$$v = -v_{max} \sin(\omega t)$$

We use this equation to find a value of ωt for which $v = -\frac{1}{2}v_{max}$. (We could have chosen $v = +\frac{1}{2}v_{max}$; the final answer would be the same.)

Thus $-\frac{1}{2}v_{max} = -v_{max}\sin(\omega t)$ so $\omega t = \sin^{-1}(\frac{1}{2}) = \frac{\pi}{6}$.

So $x = A\cos(\omega t) = 0.035\,\text{m} \times \cos\frac{\pi}{6} = 0.030\,\text{m}$

A sketch-graph of x against t shows that *whenever* $x = \pm 0.30\,\text{m}$, the ball will have half its maximum speed (gradient $= \pm\frac{1}{2}$ maximum gradient).

Grade boost

Note how the zeros in the v–t graph correspond to zero gradient in x–t, and maxima in v–t to maximum gradient in x–t. This enables you to sketch a v–t graph given an x–t graph, even if ε is not zero.

≫ Pointer

The factor of ω in $v_{max} = A\omega$ accords with the larger rate of change of x at higher frequencies. The equations quoted for v and a are obtained by successive differentiations (with respect to t) of $x = A\cos(\omega t + \varepsilon)$.

quickfire

(8) A body performs shm of period $1.2\,\text{s}$ and amplitude $0.050\,\text{m}$. Calculate:

a) the body's maximum velocity

b) the velocity $0.40\,\text{s}$ after passing through equilibrium, moving in the $+x$ direction.

Acceleration of a body in shm

The gradient of the tangent to the v–t graph at any time, t, is the acceleration at that time. Hence the shape of the a–t graph. In fact, a varies with t according to

$$a = -A\omega^2 \cos(\omega t + \varepsilon).$$

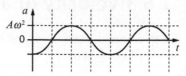

Fig. 1.6.9 a–t graph

But $\quad x = A\cos(\omega t + \varepsilon)$ So $a = -\omega^2 x$.

This backs up our earlier claim that a displacement varying as $x = A\cos(\omega t + \varepsilon)$ implies shm, as defined by $a = -\omega^2 x$ (with constant ω).

1.6.4 Energy interchange in shm

Potential energy, E_p

Grade boost

Make sure that you can sketch the $E_k - t$ and $E_p - t$ graphs.

≫ Pointer

The equation $E_p = \frac{1}{2}mA^2\omega^2$ holds even if E_p is not elastic PE, or not wholly elastic. For vertical oscillations of a body hanging from a spring, E_p is a sum of elastic and gravitational PEs.

Consider our original case: a body attached to a horizontal spring and displaced by x from its equilibrium position. The spring stores elastic PE (equal to the area under the force–extension graph), given by

$$E_p = \tfrac{1}{2}kx^2 \quad \text{which, because } \omega^2 = \frac{k}{m}, \text{ can be written} \quad E_p = \tfrac{1}{2}m\omega^2 x^2.$$

It follows that

- $E_p = 0$ whenever $x = 0$. (This is a matter of convention.)
- E_p is never negative.
- $E_p = \tfrac{1}{2}m\omega^2 A^2$ when $x = \pm A$

Hence the points on the E_p–t graph. Note that we've put $\varepsilon = 0$ in all graphs.

The shape of the graph is sinusoidal, but raised up, and of frequency $2f$.

E_p and t are related by

$$E_p = \tfrac{1}{2}m\omega^2 A^2 \cos^2(\omega t + \varepsilon)$$

(a)

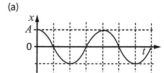

(b)

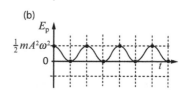

Fig. 1.6.10 Variation with time of displacement and PE

Kinetic energy, E_k

(a)
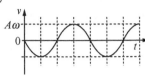

(b)
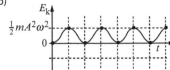

(c)
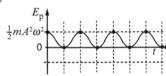

(d)
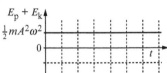

Fig. 1.6.12 Variation with time of velocity, KE, PE, total energy

The body's **KE** is $E_k = \frac{1}{2}mv^2$. So

- $E_k = 0$ whenever $v = 0$,
- E_k is never negative,
- $E_k = \frac{1}{2}m\omega^2 A^2$ when $v = \pm v_{max} = \pm A\omega$.

Hence the points plotted on the E_k–t graph (Fig. 1.6.12 (b)).

E_k and t are related by

$$E_k = \frac{1}{2}m\omega^2 A^2 \sin^2(\omega t + \varepsilon)$$

The graph is just like the E_p–t graph, but shifted along, so that E_k is high when E_p is low and vice versa.

Indeed we can show that

$$E_k + E_p = \frac{1}{2}mA^2\omega^2,$$

so the *total* energy of the system never changes (Fig. 1.6.12 (d)).

Energy is transferred from kinetic to potential and vice versa twice per cycle as the system (for example, pendulum, mass-on-spring) oscillates – a classic case of the conservation of energy.

Example

Calculate E_p and E_k for a mass–spring system ($m = 1.20\,$kg) in shm of frequency $2.5\,$Hz and amplitude $0.060\,$m, when $x = 0.030\,$m.

Answer

$E_p = \frac{1}{2}m\omega^2 x^2 = \frac{1}{2}1.20\,(2\pi \times 2.5)^2(0.030)^2\,J = 0.133\,$J

The *total* energy is equal to the *maximum* E_p, which we can find by replacing $(0.030)^2$ by $(0.060)^2$ and is therefore $4 \times 0.133\,$J$ = 0.533\,$J.

So, when $x = 0.030\,$m, $E_k = 0.533\,$J$ - 0.133\,J = 0.400\,$J

>> Pointer

Energies can also be plotted against displacement, as in Fig. 1.6.11. Check that these graphs make sense to you.

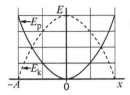

Fig. 1.6.11 Energy against displacement

quickfire

⑨ A simple pendulum bob of mass $100\,$g swings with period $2.00\,$s and amplitude $20\,$mm Calculate its PE and KE at $0.200\,$s after it has passed through vertical.

Key Terms

Natural oscillations (free oscillations) occur when an oscillatory system (such as a pendulum or a mass on a spring) is displaced and released.

The frequency of the free oscillations is called the system's **natural frequency**.

Damping is the dying away, due to resistive forces, of the amplitude of free oscillations.

Critical damping is the case when the resistive forces on the system are just large enough to prevent oscillations occurring when the system is displaced and released.

Grade boost

Before sketching damped oscillations, mark the time scale at equal intervals, and draw exponential guide-lines (dotted).

quickpire

⑩ In Fig. 1.6.13, what fraction of (a) the amplitude, (b) the total energy, at the start of a cycle remains at the end of that cycle?

1.6.5 Damped oscillations

Real **natural oscillations** (see **Key terms**) are **damped**: their amplitude decreases with time. This is due to resistive forces like air resistance, which act on the oscillating body in the opposite direction to its velocity. Shm is an extreme or ideal case: the amplitude is constant when the resistive forces are negligible.

Work done by the body against resistive forces transfers energy from the system into the random energy of the particles of the body and – in the case of air resistance – the surrounding air. This is sometimes called *dissipation* of energy. The air and the body both get slightly warmer; for the oscillating system $E_k + E_p$ decreases.

Fig. 1.6.13 is an $x-t$ graph for damped natural oscillations.

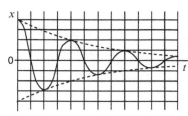

Fig. 1.6.13 Damped oscillations

- The periodic time stays the same throughout the motion.

- The amplitude **falls exponentially** with time: each cycle it decreases by the same *fraction*. (This requires the resistive force to be proportional to speed – which is often the case.)

The resistive forces increase the periodic time. The effect is very small unless the damping is very heavy. (Even in the case sketched in Fig. 1.6.13, the period is only 0.6% longer than if there were no damping.)

If damping is increased (maybe by giving the body a larger and larger surface area) we will eventually reach so-called **critical damping**: the body doesn't oscillate if displaced and released from rest, but returns towards equilibrium without overshoot (Fig. 1.6.14: middle graph). Further increase in damping results in slower return (without overshoot).

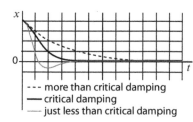

--- more than critical damping
— critical damping
— just less than critical damping

Fig. 1.6.14 Very heavy damping

Uses of critical damping

Car suspensions use springs that compress when wheels hit a ramp in the road. The car would then oscillate like an upside-down mass on a spring, but damping devices ('shock absorbers') provide (almost) critical damping, so the car returns to equilibrium height above ground quickly and with negligible overshoot.

Some door-closer mechanisms use critical damping.

1.6.6 Forced oscillations

Forced oscillations occur when a system capable of natural oscillations is subjected to a sinusoidally varying *driving force*.

The system soon settles down to oscillating at the frequency of the driving force.

Forced oscillations can be investigated with the apparatus shown on the left in Fig. 1.6.15. The *oscillatory system* is the mass-and-spring. The sinusoidally varying force is applied to it by the (constant amplitude) vibrating pin, powered by a signal generator. From this we select various frequencies of driving force, and measure the steady-state amplitude of the mass's oscillations using the scale.

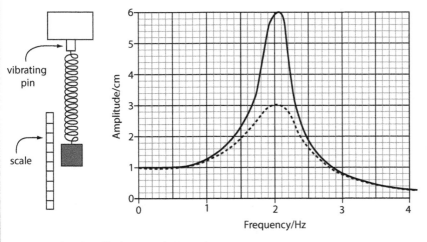

Fig. 1.6.15 Forced oscillations experiment and resonance curve

The graph (solid line) of amplitude against driving force frequency is at a maximum when this frequency is (very nearly) equal to the *natural* frequency of the mass–spring system. This peaking is called **resonance**. The curve is called a 'resonance curve' or a 'response curve'.

Increasing the damping reduces the amplitude of the forced oscillations, but reduces it by a smaller and smaller fraction the further away we go from resonance. So the curve (broken line) is not as *sharp*. The peak occurs at a slightly lower frequency of driving force.

An off-resonance use of forced oscillations

In a microwave oven, water (and some of the fats) in food absorb energy from microwaves. The alternating electric field in the waves exerts forces on the electrons and protons in the molecules, making the molecules rotate to and fro – an oscillating motion. There is heavy damping due to interactions with neighbouring molecules, and energy becomes randomly distributed, as shown by a rise in temperature. The frequency of the microwaves (2.45 GHz) is well away from any natural frequency of the molecules.

Grade boost

Make sure you know – and understand – the definitions of natural (free) and forced oscillations.

quickfire

⑪ What is the *natural* frequency of the mass-spring system in Fig. 1.6.15?

Grade boost

Make sure you can sketch a resonance curve. Note that however low the forcing frequency, the system will still oscillate, so there will be a finite amplitude.

A use of resonance

Certain electrical circuits have natural frequencies of oscillation of current through them. When you select (or 'tune in') a particular radio station, you are adjusting the natural frequency of one of these circuits to that of the radio waves from the station, so its waves force a much larger current through the circuit than other stations' waves.

When resonance must be avoided or suppressed

A suspension bridge has various modes of natural oscillation. When people walk over it, their footsteps exert periodic driving forces. If the frequency of these forces is close to a natural frequency of the bridge, the structure can make forced oscillations of alarmingly high amplitude. As found on the opening day of London's *Millennium Bridge*, people tend (probably not deliberately) to fall into step with any incipient swaying of the bridge, so ensuring resonance! The problem was solved by adding *damping* mechanisms to the bridge.

>> **Pointer**

Resonance can be a nuisance even when not dangerous. Rattles or buzzings can occur when the rotation rate of machines touches on the natural frequency of an item that can vibrate. Old vehicles may exhibit such resonances. More modern vehicles contain more damping material!

1.6.7 Specified practical work

(a) Determining g with a simple pendulum

With very basic apparatus (Fig. 1.6.16) and some care, g can be found with an uncertainty of less than 5%.

We need to determine the periodic time, T, for a known length, l, of pendulum.

To check the equation $T = 2\pi\sqrt{\dfrac{l}{g}}$ we repeat this for a range of values of l, typically about from 10 cm to 100 cm.

The string length, s (Fig. 1.6.16), is measured with a metre rule, and the diameter, d, of the bob with Vernier callipers. Clearly

$$l = s + \frac{d}{2}$$

T is found by timing n cycles of oscillation in one go and dividing by n.

- The angle to the vertical shouldn't exceed about 20°.
- The bob spends a relatively long time near the extremes of its swing, making it difficult to judge when it's *exactly* there. That's why we usually start and finish timing with the bob crossing the equilibrium position (and moving the same way each time). It helps to use a *fiducial mark* (Fig. 1.6.16).
- n should be chosen so that the uncertainty in T is less than about 2%. If you want to time more than 50 swings or fewer than 5, you've probably misjudged!

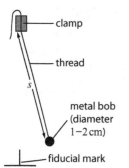

clamp

thread

s

metal bob
(diameter
1–2 cm)

fiducial mark

Fig. 1.6.16 Simple pendulum

The relationship between T and l given above can be written,

$$T = \frac{2\pi}{\sqrt{g}}\sqrt{l}$$

Comparing with $y = mx + c$, we see that plotting T against $\sqrt{l}$ (see **Grade boost**) should give a straight line of gradient $\frac{2\pi}{\sqrt{g}}$ through the origin. Hence a value for g.

A neat alternative is *not* to add $\frac{d}{2}$ to each value of s, but to plot s itself against T^2. The gradient will be $\frac{g}{4\pi^2}$ and the intercept, $\frac{d}{2}$. See Quickfire 12.

Hence a value for g, and – with a horrendous percentage uncertainty – for the bob diameter, d.

(b) Investigating damping in a mass–spring system.

We use the apparatus in Fig. 1.6.3 with a ruler clamped vertically, close to the up and down oscillations of the mass. The period should be more than about 1.5 s so that the oscillations can be followed clearly. (See first **Pointer**). Once set, T should be found accurately by timing many cycles in one go.

The equilibrium position of a point, **P**, on the mass (or a pointer fixed to the mass) is read on the ruler. Beware parallax! Oscillations are started by pulling down the mass and releasing it, noting **P**'s extreme position on the ruler initially, 5 cycles later, 10 cycles later and so on (See **Pointer**) Hence we can determine the amplitude initially and at intervals of five cycles.

The amplitude, A, is expected to vary with time, t, according to

$$A = A_0 e^{-\lambda t}$$

in which A_0 is the initial amplitude (the value of A when $t = 0$), and λ is a constant called the *damping coefficient*.

After n cycles, $t = Tn$, in which T is the period.

So
$$A = A_0 e^{-\lambda Tn}$$

To obtain a straight-line graph relationship we take logarithms (to any base) of both sides. Choosing the base e, we have

$$\ln(A/\text{mm}) = \ln(A_0/\text{mm}) - \lambda Tn$$

- A and A_0 stand for numbers multiplied by units (e.g. 15 mm). A/mm and A_0/mm, though, are pure numbers; for example, if $A = 15\,\text{mm}$, then $A/\text{mm} = 15\,\text{mm/mm} = 15$. Only pure (positive) numbers can have logarithms. Even the most expensive calculator won't give you a value for $\ln(15\,\text{mm})$!

Comparing the last equation with $y = mx + c$, we see that $\ln(A/\text{mm})$ against n should be a straight line of gradient λT. Hence λ, knowing T.

You may prefer to plot $\ln(A/\text{mm})$ against t, calculating t for each value of n, from $t = Tn$.

quickfire

⑫ Justify the 'neat alternative' given in the main text, by showing that $s = \frac{g}{4\pi^2}T^2 - \frac{d}{2}$ Start by squaring both sides of $T = 2\pi\sqrt{\dfrac{l}{g}}$.

quickfire

⑬ At the trying-out stage, what should be done to increase the oscillation period if it is too short?

》 Pointer

Why not determine the amplitude at intervals of a single cycle? It may not be possible to record the results fast enough. But there's nothing magic about a 5-cycle interval.

》 Pointer

A nice way to extend investigation (b) is to repeat it with more damping, for example with a disc of paper glued underneath the mass so that it sticks out by a few cm all round. In the version with less damping the paper should, ideally, be folded up small and attached to the mass. Why?

1. A mass of 0.20 kg hangs from a spring clamped at the top. The mass is pulled down from its equilibrium position and released. A graph of acceleration against displacement from equilibrium is given for its motion after release.

    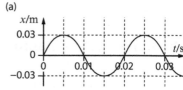
    Fig. 1.6.17 a–x graph

    (a) Explain how the graph in Fig 1.6.17 shows the motion of the mass to be simple harmonic.

    (b) Calculate the spring constant, k.

    (c) Calculate the periodic time of the motion.

    (d) Calculate how long the mass will take to **rise 0.020 m from its point of release**. [Hint: you'll need another piece of data from the graph.]

2. An x–t graph is given in Fig. 1.6.18 (a) for a mass–spring system.

    (a) Determine the frequency of the oscillations.

    (b) Copy the grid in Fig. 1.6.18 (b) and sketch on it a velocity–time graph for the mass, putting the value of the maximum velocity on the v axis.

    (c) Calculate the mass's velocity at $t = 0.003$ s, stating the next time when this velocity occurs.

    (d) Calculate the velocity when $x = 0.020$ m.

    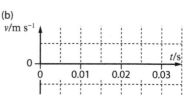
    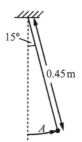
    Fig. 1.6.18 x–t graph and v–t grid

3. A simple pendulum of length 0.450 m has a bob of mass 0.060 kg. The pendulum is released from an angle of 15° to the vertical (Fig 1.6.19).

    (a) Show that the height lost by the bob as it swings to its lowest point is $(1 - \cos 15°)\,0.450$ m.

    (b) Hence, using the principle of conservation of energy, calculate the bob's maximum kinetic energy.

    (c) Show that the angle of 15° corresponds to the shm having an amplitude, A, along the arc (Fig. 1.6.19) of 0.118 m. Revise Section 1.5 if necessary!

    (d) Determine the maximum velocity of the pendulum using the relevant shm equation. (You will need to calculate the pendulum's period, or equivalent.)

    (e) Use (d) to recalculate the pendulum's maximum KE.

    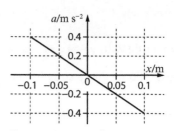
    Fig. 1.6.19 Initial pendulum

4. A student has timed 30 cycles of a simple pendulum (with repetition) for the shortest and longest lengths of a pendulum he intends to use. His readings are:

string length / m	time for 30 cycles / s
0.150 ± 0.001	23.9 ± 0.2
0.750 ± 0.001	52.4 ± 0.2

    The student runs out of time and cannot take more readings. He decides to determine the best value for g that he can from his results. He plots the two points, with error bars, on a graph of T^2 against string length, s, and proceeds from there. Determine:

    (a) the maximum and minimum gradients consistent with his data.

    (b) the value of g that this gives, together with its absolute uncertainty.

1.7 Kinetic theory of gases

Although we can't see what's going on inside a gas, using either optical or non-optical microscopes, a 'model' of a gas as particles (molecules) in random motion successfully explains the properties we observe. In particular, applying Newton's laws to the molecules accounts for gas *pressure* and how it depends on the container volume and on the temperature.

1.7.1 Molecules and moles

Gases consist of particles called **molecules**. In so-called **monatomic** gases (such as helium and neon) the molecules are single atoms. In hydrogen, oxygen, nitrogen and many other gases, each molecule consists of more than one atom 'bonded' together.

The **relative molecular mass**, M_r, of a molecule is defined as

$$M_r = \frac{\text{mass of molecule}}{\frac{1}{12}\text{mass of }^{12}_{6}\text{C atom}}$$

M_r examples: hydrogen: 1.01, helium: 4.00, oxygen: 32.0, nitrogen: 28.0. (You don't have to remember these!)

Moles

A mole (abbreviates to 'mol') of molecules is a batch of 6.02×10^{23} molecules. The number of molecules per mole is called the **Avogadro constant**, N_A.

Thus $N_A = 6.02 \times 10^{23}\text{mol}^{-1}$.

It follows that, if we have N molecules, the **amount**, n in moles, is:

$$n = \frac{N}{N_A} \quad \text{so} \quad N = nN_A$$

The **molar mass**, M, is the mass per mole of the gas...

There is an easy relationship between the M_r of a molecule and the molar mass, M:

$$M/\text{kg mol}^{-1} = \frac{M_r}{1000}$$

Example

How many molecules are there in 20.0 kg of oxygen gas?

Answer

molar mass $= \dfrac{M_r}{1000}\text{kg mol}^{-1} = \dfrac{32.0}{1000}\text{kg mol}^{-1} = 0.0320\text{ kg mol}^{-1}$

So, amount in moles, $n = \dfrac{\text{mass of gas}}{\text{molar mass}} = \dfrac{20.0\text{ kg}}{0.0320\text{ kg mol}^{-1}} = 625$ mol

and number of molecules, $N = nN_A = 625\text{ mol} \times 6.02 \times 10^{23}\text{ mol}^{-1}$

$\qquad = 3.76 \times 10^{26}$

>> **Pointer**
N_A is not arbitrary, but is chosen to give the simple equation
$$M = \frac{M_r}{1000}\text{ kg mol}^{-1}$$

Grade boost
Try to remember what n, N, M, M_r stand for: they're fairly standard.

>> **Pointer**
Remember
$$n = \frac{\text{Total mass}}{\text{molar mass}}$$

quickpire
① Calculate the mass (in kg) of an oxygen molecule ($M_r = 32.0$).

quickpire
② What is the mass of 1.00×10^{25} helium molecules?
($M_r = 4.00$)

>> *Pointer*

The kelvin (K) is the SI unit of temperature. Note the plain 'K'. Contrast with °C for degree Celsius.
Note that
$T/K = \theta/°C + 273.15$

Key Term

Absolute zero, 0 K, is the temperature at which the energy of the particles of bodies is as low as it possibly can be.

quickfire

③ For the sample of gas for which the graphs are drawn, calculate the pressure at 300 K when the volume is $0.090 \times 10^{-3}\,m^3$.
(Use $pV = $ constant.)

quickfire

④ The solid line and the broken line in Fig. 1.7.2 are both for the same sample of gas. What is the gas temperature when the broken line applies?

1.7.2 Gas pressure: experimental

A gas exerts a force on any surface it's in contact with – at right angles to the surface. The **pressure** is defined as the magnitude of that force per unit area of surface. It is a scalar.

(a) Dependence on container volume

A sample of gas can be squeezed into a (much) smaller volume. The smaller the volume, the larger the pressure the gas exerts. More precisely, we find...

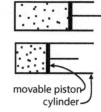

movable piston
cylinder

Fig. 1.7.1 Squeezing a gas

Boyle's law

For a fixed mass of gas at constant temperature, the pressure (p) it exerts is inversely proportional to the volume (V) it occupies.

This means that if we halve V, then p doubles; if we treble V then p decreases by a factor of 3, and so on. (Check either graph!) This is equivalent to

$$pV = \text{constant}$$

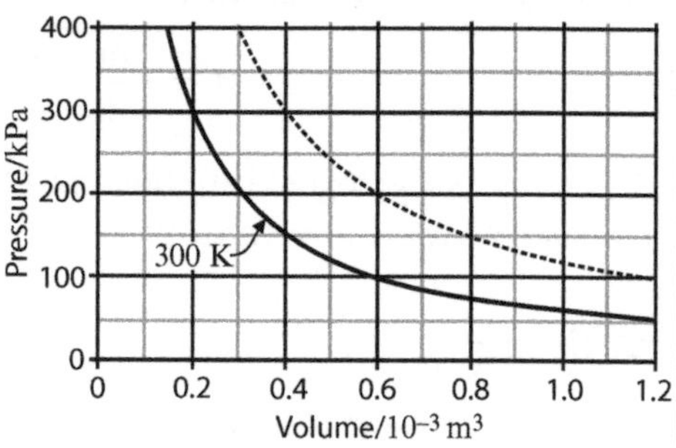

Fig. 1.7.2 Boyle's law

provided that temperature and amount of gas don't change.

Boyle's law is an experimental law. It applies quite well to gases at 'ordinary' densities, but with greater and greater accuracy as the gas density approaches zero (when mean separation of molecules is increased). We say that the gas is approaching **ideal gas** behaviour.

(b) Dependence on temperature

The higher the temperature, the larger the value of pV for a sample of gas. In fact, for an ideal gas, pV is *proportional* to temperature, T, measured on the so-called kelvin scale. This is not just luck: T can be *defined* as proportional to pV for a sample of gas (as its density approaches zero)!

T, so defined, is an **absolute temperature**: it has its zero at **absolute zero**: the temperature at which an ideal gas would have $pV = 0$. This is the lowest temperature possible, the temperature at which the particles of matter have their least possible random energy.

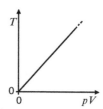

Fig. 1.7.3 $pV \propto T$

The details (beyond A-Level!) of the definition of the **kelvin** scale imply that the freezing point and boiling point of water (at a pressure of 101.3 kPa) are, to 4 significant figures, 273.2 K and 373.2 K.

Temperatures in degree Celsius are now *defined* as

$$\theta/°C = T/K - 273.15, \quad \text{that is} \quad T/K = \theta/°C + 273.15.$$

(c) Dependence on number of molecules

For given values of V and T, the pressure is proportional to N, the number of molecules (or, equivalently, to n, the number of moles).

(d)The ideal gas equation

(a), (b) and (c) above are all included in the **ideal gas equation** ...

$$pV = nRT \quad \text{or, equivalently} \quad pV = NkT$$

n is the amount of gas in moles. $\quad N$ is the number of molecules.
R is the molar gas constant. $\quad k$ is the Boltzmann constant.
$R = 8.31\,\text{J}\,\text{mol}^{-1}\,\text{K}^{-1}$. $\quad k = 1.38 \times 10^{-23}\,\text{J}\,\text{K}^{-1}$

Clearly $Nk = nR \quad$ so $\quad Nk = \dfrac{N}{N_A}R \quad$ so $\quad k = \dfrac{R}{N_A}$

Note how the ideal gas equation contains Boyle's law: when n (or N) and T are constant, nRT (or NkT) is constant, so pV is constant.

Example

A car tyre contains $0.0140\,\text{m}^3$ of nitrogen at a pressure of 320 kPa and a temperature of 290 K. Calculate the number of nitrogen molecules and the mass of nitrogen.

Answer

$$N = \frac{pV}{kT} = \frac{320 \times 10^3 \times 0.014\,\text{J}}{1.38 \times 10^{-23}\,\text{J}\,\text{K}^{-1} \times 290\,\text{K}} = 1.12 \times 10^{24}$$

$$n = \frac{N}{N_A} = \frac{1.12 \times 10^{24}}{6.02 \times 10^{23}\,\text{mol}^{-1}} = 1.86\,\text{mol};$$

$$\text{molar mass of nitrogen} = \frac{28.0}{1000}\,\text{kg}\,\text{mol}^{-1}$$

$$\text{so mass of nitrogen} = 1.86\,\text{mol} \times \frac{28.0}{1000}\,\text{kg}\,\text{mol}^{-1} = 0.052\,\text{kg}$$

Pointer

The units of pV are $\text{N}\,\text{m}^{-2} \times \text{m}^3 = \text{Nm} = \text{J}$. Check the units given for R and k.

quickfire

⑤ For the gas sample whose p–V graphs are plotted in Fig. 1.7.2, calculate the amount in moles and the number of molecules.

quickfire

⑥ Calculate the gas pressure in the tyre in the example, if its temperature rises to 300 K, and its volume rises to $0.0142\,\text{m}^3$. [Assume no gas escapes.]

Pointer

Key equations are
$pV = nRT$
$pV = NkT$
$k = \dfrac{R}{N_A}$

Although we can't see individual moving molecules, we can see with a microscope the slower 'Brownian movement' of larger particles buffeted by gas molecules.

>> **Pointer**

Check that the units are the same on both sides of the equations:

$$p = \tfrac{1}{3}\rho\overline{c^2} = \tfrac{1}{3}\frac{N}{V}m\overline{c^2}$$

 Grade boost

Make sure you know what m and N stand for in $pV = \tfrac{1}{3}Nm\overline{c^2}$.

1.7.3 The kinetic theory of an ideal gas

A gas consists of particles (called molecules) in rapid random motion in space which is otherwise empty. The molecules continually collide with each other and with the walls of the container. We assume that:

- Collisions between molecules are elastic (at least on average).
- The molecules themselves occupy a negligible fraction of the container volume.
- The molecules exert negligible forces on each other, and move in straight lines, except when colliding.

The last two assumptions become more and more realistic the lower the gas density, that is as the gas approaches an ideal gas.

Individual molecules moving in a gas cannot be seen, even with special microscopes, but the theory's success lies in what it can explain. Start with how a gas can be squeezed into a smaller volume: according to the kinetic theory we are simply reducing empty space between molecules!

(a) Kinetic theory of gas pressure

Gas pressure is caused by the random bombardment of the container walls by molecules. Because of the randomness, the pressure will be the same on any 'patch' of the container wall, and given by

$$\text{pressure} = \frac{\text{magnitude of mean normal force on wall}}{\text{area of wall over which force acts}}$$

The force due to the random hits is equal and opposite to the mean rate of change of the molecules' momenta as they collide with the wall and bounce back. Here we are using Newton's second and third laws. Working out the consequences mathematically leads to the equation...

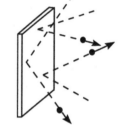

Fig. 1.7.4 wall hit by molecules

$$pV = \tfrac{1}{3}Nm\overline{c^2} \quad \text{or its alternative form,} \quad p = \tfrac{1}{3}\rho\overline{c^2}.$$

In these equations, p and V are gas pressure and container volume.

N is the number of molecules in the container.

m is the mass of one molecule.

ρ is the gas density. Because Nm is the total mass of all the molecules in volume V, then $\rho = Nm/V$, hence the alternative form of the kinetic theory equation.

$\overline{c^2}$ (or c_{rms}) is the **mean square speed** of the molecules, defined by

$$\overline{c^2} = \frac{c_1{}^2 + c_2{}^2 \ldots\ldots + c_N{}^2}{N}$$

in which $c_1, c_2 \ldots\ldots c_N$ are the speeds of the individual molecules.

There will be a wide range of speeds, as some molecules will have acquired high speeds in 'lucky' collisions, others much lower.

We define the **root mean square (rms) speed, c_{rms},** by

$$c_{rms} = \sqrt{\overline{c^2}} = \sqrt{\frac{c_1^2 + c_2^2 \ldots\ldots + c_N^2}{N}}$$

The significance of c_{rms} is that if all the molecules had this one speed, the gas pressure would be the same as for the actual speeds ($c_1, \ldots c_N$).

Example

Calculate c_{rms} for 3 molecules of speeds $357\,\text{m s}^{-1}$, $401\,\text{m s}^{-1}$, $532\,\text{m s}^{-1}$.

Answer

$$\overline{c^2} = \frac{357^2 + 401^2 + 532^2}{3} = 1.90 \times 10^5 \text{ m}^2\,\text{s}^{-2}; \quad c_{rms} = \sqrt{\overline{c^2}} = 436 \text{ m s}^{-1}.$$

(b) Finding the rms speed of molecules

The equation $p = \frac{1}{3}\rho\overline{c^2}$ enables us to find the rms speed of the molecules – even though we can't see them – from easily measurable 'everyday' quantities.

Example

At room temperature and a pressure of 101 kPa, $1.00 \times 10^{-3}\,\text{m}^{-3}$ of nitrogen is found to have a mass of 1.17×10^{-3} kg. Determine c_{rms}.

Answer

$$\rho = \frac{\text{mass}}{\text{volume}} = \frac{1.17 \times 10^{-3} \text{ kg}}{1.00 \times 10^{-3} \text{ m}^3} = 1.17 \text{ kg m}^{-3}$$

$$c_{rms} = \sqrt{\overline{c^2}} = \sqrt{\frac{3p}{\rho}} = \sqrt{\frac{3 \times 101 \times 10^3 \text{ N m}^{-2}}{1.17 \text{ kg m}^{-3}}} = 509 \text{ m s}^{-1}$$

1.7.4 The mean kinetic energy of molecules

We have equations for pV in terms of molecular quantities, and in terms of temperature:

$$pV = \frac{1}{3}Nm\overline{c^2} \quad \text{and} \quad pV = NkT.$$

Equating the two right-hand expressions for pV...

$$\frac{1}{3}Nm\overline{c^2} = NkT \quad \text{so} \quad \frac{1}{3}m\overline{c^2} = kT$$

so finally $\qquad \frac{1}{2}m\overline{c^2} = \frac{3}{2}kT$

$\frac{1}{2}m\overline{c^2}$ is the mean *translational* kinetic energy of a gas molecule, that is the **KE** of its moving around (as opposed to **KE** of rotation, which some molecules also have). As we see, it is proportional to T. So *kelvin temperature is a measure of the mean translational* **KE** *of a gas molecule.*

❱❱ Pointer

c_{rms} is greater than the mean speed, $\overline{c}$ (unless all the molecules have the same speed!).

quicKfire

⑦ Calculate c_{rms} for two molecules, with speeds 200 m s^{-1}, 400 m s^{-1}.

Grade boost

To find c_{rms} it's usually easier to use $p = \frac{1}{3}\rho\overline{c^2}$ than $pV = \frac{1}{3}Nm\overline{c^2}$.

Sometimes (as in part (b) of the first example in 1.7.4) you may not have the data to use either. But at least in that example, part (a) gave a hint!

Grade boost

Make sure you can derive $\frac{1}{3}m\overline{c^2} = \frac{3}{2}kT$.

It's also worth remembering it for quick recall.

quicKfire

⑧ A cylinder of volume $0.025\,\text{m}^3$ contains 6.1 mol of oxygen ($M_r = 32.0$) at a pressure of 600 kPa. Calculate:

a) the mass of gas
b) the density
c) the molecules' rms speed
d) the temperature
e) the molecules' mean translational KE.

⑨ Determine the rms speed of oxygen molecules in the air under conditions when the rms speed of the nitrogen molecules is $500\,\text{m s}^{-1}$.

Example

Calculate (a) the mean translational kinetic energy (b) the rms speed of a nitrogen molecule ($M_r = 28.0$), at 300 K.

Answer

(a) $\frac{1}{2}m\overline{c^2} = \frac{3}{2}kT = \frac{3}{2} \times 1.38 \times 10^{-23}\,\text{J K}^{-1} \times 300\,\text{K} = 6.21 \times 10^{-21}\,\text{J}$

(b) $m = \dfrac{\text{molar mass}}{N_A} = \dfrac{0.028\,\text{kg mol}^{-1}}{6.02 \times 10^{23}\,\text{mol}^{-1}} = 4.65 \times 10^{-26}\,\text{kg}$

But $\frac{1}{2}m\overline{c^2} = 6.21 \times 10^{-21}\,\text{J}$ so $\overline{c^2} = \dfrac{2 \times 6.21 \times 10^{-21}\,\text{J}}{4.65 \times 10^{-26}\,\text{kg}}$

And $c_{\text{rms}} = \sqrt{\overline{c^2}} = \sqrt{\dfrac{2 \times 6.21 \times 10^{-21}\,\text{J}}{4.65 \times 10^{-26}\,\text{Kg}}} = 517\,\text{m s}^{-1}$

Molecules of different gases at the same temperature

According to the equation, $\frac{1}{2}m\overline{c^2} = \frac{3}{2}kT$, all gas molecules at the same temperature have the same mean translational kinetic energy. This implies that, at the same temperature, molecules of greater mass have smaller rms speeds.

Example

Show that in a mixture of hydrogen gas and oxygen gas, the rms speed of the hydrogen molecules is nearly four times that of the oxygen molecules. (M_r: hydrogen (H_2): 2.02, oxygen (O_2): 32.0.)

Answer

We may assume that in a mixture of gases, the components will be in thermal equilibrium (Section 1.8.3 (b)) and at the same temperature.

So $\frac{1}{2}m_{H2}\overline{c_{H2}}^2 = \frac{1}{2}m_{O2}\overline{c_{O2}}^2$

Therefore $\dfrac{\overline{c_{H2}}^2}{\overline{c_{O2}}^2} = \dfrac{m_{O2}}{m_{H2}} = \dfrac{32.0}{2.02}$ so $\dfrac{c_{\text{rms H2}}}{c_{\text{rms O2}}} = \sqrt{\dfrac{\overline{c_{H2}}^2}{\overline{c_{O2}}^2}} = \sqrt{\dfrac{32.0}{2.02}} = 3.98$

Just make sure there are no sparks!

1. Show that the ideal gas equation implies that equal volumes of gases (even if they are different gases) at the same temperature and pressure contain equal numbers of molecules. (This is called *Avogadro's law*.)

2. (a) Suppose that a gas molecule of mass m hits a wall of the gas container with a velocity component u normal to the wall, and that it bounces off the wall with normal velocity component $-u$. Use Newton's second and third laws to show clearly that if there are f such hits with the same wall area, A, per unit time, the mean normal force on it is $2fmu$. [Further development – beyond A-level – of this argument gives us $p = \frac{1}{3}\rho\overline{c^2}$.]

   (b) Suggest *two* reasons, based on (a), for a gas exerting more pressure on the walls of its container if the mean speed of the molecules rises. Hint: don't forget f.

3. 0.050 mole of helium gas is contained in a cylinder fitted with a leak-proof piston. The piston moves slowly outwards resulting in the pressure variation in Fig. 1.7.5.

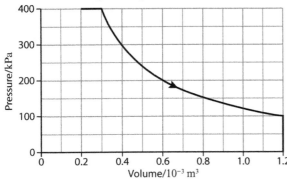

Fig. 1.7.5 p–V graph

   (a) Verify that the gas temperature does not change, and calculate this temperature.

   (b) Calculate the mean (translational) energy of a molecule of the gas.

   (c) Calculate the rms speed of the molecules (M_r of helium = 4.00).

4. (a) In a cylinder of helium gas, three of the molecules have speeds of 950 m s^{-1}, 1300 m s^{-1} and 1650 m s^{-1} at one instant. Calculate the rms speed of this group of molecules at that instant.

   (b) Explain why the speed of a gas molecule frequently changes.

   (c) The cylinder has a volume of 5.0×10^{-3} m^3 and contains 8.0×10^{-3} kg of helium gas at a pressure of 900 kPa. Calculate:

   (i) the number of helium molecules ($M_r = 4.00$) in the cylinder

   (ii) the rms speed of all the molecules in the cylinder

   (iii) the rms speed of all the molecules if the kelvin temperature were doubled.

1.8 Thermal physics

We'll be concerned mainly with thermodynamics – studying the internal energy of a system and transfers of energy as work and heat into and out of the system. The laws of thermodynamics apply to systems ranging from rubber bands to stars. The favourite system at A-level is a sample of ideal gas.

1.8.1 Internal energy

This is defined in **Key terms**. An important case is…

The internal energy of an ideal monatomic gas

An ideal gas has negligible forces between its molecules, except during collisions, so the potential energy of interactions can be taken as zero.

So U = sum of kinetic energies of N molecules

$$= N \times \text{mean kinetic energy of a molecule} = N \times \tfrac{1}{2}m\overline{c^2}$$

But (see Section 1.7.4) $\tfrac{1}{2}m\overline{c^2} = \tfrac{3}{2}kT$

So $U = \tfrac{3}{2}NkT$ or, since $N = nN_A$, $k = \dfrac{R}{N_A}$, $U = \tfrac{3}{2}nRT$

This equation applies to monatomic gases such as helium, whose molecules have only translational kinetic energy.

For other gases the $\tfrac{3}{2}$ has to be replaced by a larger factor; but, for *all* gases behaving ideally, U depends only on T.

1.8.2 p–V diagrams for a gas

We shall be studying *changes* to a system consisting of some gas in a cylinder with a piston that can be moved or held still. The temperature can be altered. The p–V diagram shows three kinds of change…

- **AB**: *change at constant volume*: Piston is held in one place; T must increase, for p to increase.

- **CD**: *change at constant pressure*: Piston moves out against constant opposing force; T must increase, for V to increase.

- **EF**: *change at constant temperature* (*a so-called* **isothermal** *change*). Check that pV is constant!

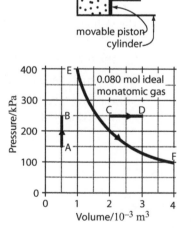

Fig. 1.8.1 (a) Gas in cylinder (b) p–V diagram

Example

Calculate the internal energy change for the expansion CD in Fig. 1.8.1.

Answer

$$T_D = \frac{P_D V_D}{nR} = \frac{250 \times 10^3 \,\mathrm{N\,m^{-2}} \times 3.00 \times 10^{-3}\,\mathrm{m^3}}{0.080\ \mathrm{mol} \times 8.31\ \mathrm{J\,mol^{-1}\,K^{-1}}} = 1130\,\mathrm{K};\ \text{similarly}\ T_C = 750\,\mathrm{K}$$

So $\quad \Delta U = U_D - U_C = \frac{3}{2}nR(T_D - T_C) = 380\,\mathrm{J}$ (see **Pointer**)

1.8.3 Energy flows into and out of a system

We classify these flows into just two kinds: *heat* and *work*.

(a) Work

A system does work $F\,\Delta x$ when it exerts a force F that moves a distance Δx in the same direction as the force. That is...

Work done by system, $W = F\,\Delta x$.

The system we're most concerned with is a gas at pressure p in a cylinder. In Fig. 1.8.2, if the piston has an area A, and moves outwards to expand the gas by a *small* volume, ΔV, then $F = pA$ and $\Delta V = A\Delta x$. So...

Work done by gas, $W = F\,\Delta x = p\,\Delta V$.

Fig. 1.8.2 Expanding gas

In a *large* expansion, p may change greatly. We must then sum all the bits of work, $p\,\Delta V$, that is the areas of all the narrow strips under the p–V graph (Fig. 1.8.3), from initial volume V_1 to final volume, V_2. So...

Work done by gas, $W = $ Area under p–V graph.

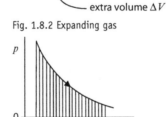

Fig. 1.8.3 'large' expansion

Examples

In Fig. 1.8.4 calculate the work done by the gas in the changes AB, CD, EF (discussed in Section 1.8.2).

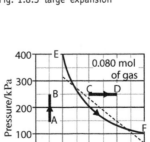

Fig. 1.8.4 p, V changes

Answers

- AB: $\Delta V = 0$ so $W = 0$. Easy one!
- CD: As the pressure is constant,
  $$W = p\,\Delta V$$
  $$= 250 \times 10^3\,\mathrm{N\,m^{-2}} \times (3.0 - 2.0) \times 10^{-3}\,\mathrm{m^3}$$
  $$= 250\,\mathrm{J}\ (\text{gas does work})$$

≫ Pointer

A neater way of doing the example (and Quickfire 1) is to write $U = \frac{3}{2}nRT$ as $U = \frac{3}{2}pV$. This saves calculating T.

≫ Pointer

For *all* changes, work done = area under p–V graph.

≫ Pointer

If the gas volume contracts (so the line on the p–V graph has a right to left trend) work is done on the gas, so W is negative.

⤊ Grade boost

If you're asked to find the work done during a change, you must state whether it is done *on* or *by* the gas.

quickfire

③ Calculate the work done when:

a) a gas contracts in volume from $0.50 \times 10^{-3}\,\mathrm{m^3}$ to $0.35 \times 10^{-3}\,\mathrm{m^3}$ at a constant p of $70 \times 10^3\,\mathrm{Pa}$.

b) the gas in Fig. 1.8.4 expands along EF, from $1.0 \times 10^{-3}\,\mathrm{m^3}$ but only to $3.0 \times 10^{-3}\,\mathrm{m^3}$.

Grade boost

Don't confuse *heat* and *temperature*! It would be like confusing electric charge and electric potential.

Grade boost

Systems don't *store* or *possess* heat. They possess *internal energy*. Heat, like work, is energy *in transit*.

quickfire

④ 0.050 mol of ideal monatomic gas expands at a constant pressure of 100 kPa from a volume of $1.25 \times 10^{-3}\,m^3$ to $1.50 \times 10^{-3}\,m^3$. Calculate:

a) the change in temperature

b) the change in internal energy

c) the work done

d) the heat flow.

- EF: We need to find the area under the curve EF. One method is to draw a straight line (the broken line) by *eye* so that the area under it is equal to that under the curve. The area under the broken line is that of a triangle sitting on a rectangle.

area = $\frac{1}{2}$triangle base × triangle height + rectangle base × rectangle height

$= \frac{1}{2}(4.0 - 1.0) \times 10^{-3} \times (310 - 60) \times 10^3\,J$
$\qquad\qquad + (4.0 - 1.0) \times 10^{-3} \times 60 \times 10^3\,J$

$= 375\,J + 180\,J.$ So work done by gas over EF = 560 J (2 sf)

(b) Heat

Heat is defined in **Key terms**. The definition covers energy transfer by conduction, convection and radiation. We'll be concerned mainly with conduction (through container walls).

Fig. 1.8.5 compares heat flow due to a temperature difference with *charge* flow due to a potential difference.

It takes time for a given amount of heat to flow (though the greater the temperature gradient, $(T_1 - T_2)/L$, the greater the rate of flow).

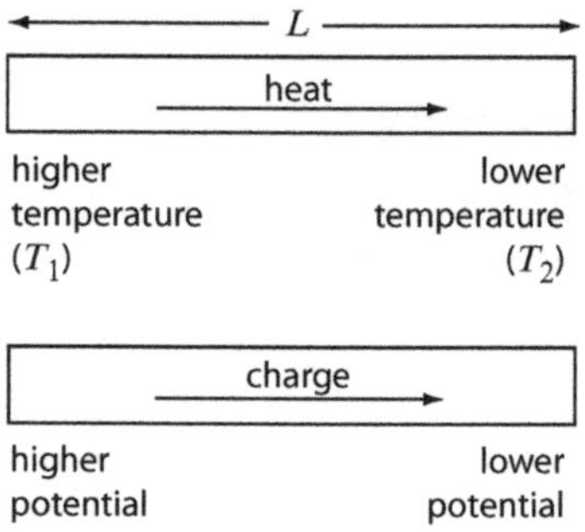

When two bodies or two regions (not thermally insulated from each other) are in **thermal equilibrium**, that is no net heat flows between them, they must be at the same temperature.

For a *system* at a lower temperature than its surroundings, energy will flow in through the system's boundary. This energy flow is *heat*, Q. For example, heat will enter a cool gas through the walls of its container if we put a flame underneath! Conversely, heat will *leave* the system if its surroundings are cooler.

Fig. 1.8.5 Heat and charge flow

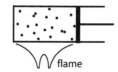

Fig. 1.8.6 Heat Input ($Q > 0$)

1.8.4 The first law of thermodynamics

$$\Delta U = Q - W$$

increase in system's internal energy   heat flow into system   work done by system

- This is a case of the principle of conservation of energy.
- ΔU is a change in a *property* of the system: internal energy, U. Q and W are *not* properties, nor changes in properties, of the system. They are both energy *in transit* between the system and the outside world.
- If heat flows *out* of the system, Q is negative. If work is done *on* the system W is negative.

Example

During some process a system has 600 J of work done on it, and its internal energy falls by 900 J. Calculate the heat flow.

Answer

$Q = \Delta U + W = (-900\text{ J}) + (-600\text{ J}) = -1500\text{ J}$

Therefore 1500 J of heat flows out of the system.

Two applications of the first law of thermodynamics

(a) Rapid expansion of an ideal gas

The significance of a *rapid* change is that it doesn't allow time for much heat to flow. $Q = 0$ (almost), so, using the *first law*:

$$\Delta U = -W.$$

Since the gas does work, W is positive, so U falls, and with it, the temperature. We can feel the cooling of air squashed into a plastic syringe, when we let the piston move out quickly, doing work against one's hand (not just letting go).

Note that the p–V graph (Fig. 1.8.7) does not obey pV = constant. Check this!

Rapid *compression* of a gas produces a temperature *rise*. In a diesel engine, the temperature rise produced by rapid compression of air is enough to ignite injected fuel (with no spark needed).

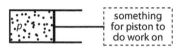

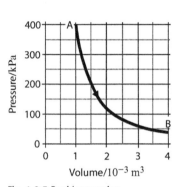

Fig. 1.8.7 Rapid expansion

(b) Slow (isothermal) expansion of an ideal gas

This time heat *will* have time to flow in. (It helps if the cylinder walls conduct heat well.) The gas temperature will not be able to fall more than slightly below that of its surroundings. Thus the expansion is essentially *isothermal*: constant temperature, so for an ideal gas, $\Delta U = 0$. Using the *first law*:

$$0 = Q - W \quad \text{that is} \quad Q = W.$$

- In this case, heat flowing into a system doesn't make it hotter!

- The isothermal expansion turns heat into work, which can, in principle, be used to lift weights, generate electricity, propel a vehicle and so on. For this transfer of energy not to be just a 'one-off', we need to take the gas over and over again through a *cycle* of changes (see next section).

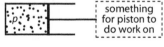

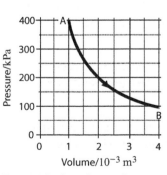

Fig. 1.8.8 Isothermal expansion

quicKfire

⑤ This question refers to Fig. 1.8.7. Calculate:
a) the temperature at B, given that it is 300 K at A
b) the change in internal energy, assuming $U = \frac{3}{2}nRT$
c) hence the work done by the gas from A to B.

quicKfire

⑥ Use the 'area' method in Section 1.8.3 (a) to check your answer to Quickfire 5 (c).

quicKfire

⑦ How, briefly, would you determine from Fig. 1.8.8 the heat flow into the gas in the isothermal case?

1.8.5 Cycles of changes

ABCDA in Fig. 1.8.9 represents some clockwise cycle of changes for a gas, with the extremes of volume at A and C.

- Over a cycle the gas returns to its original P, V and (therefore) T.

So $\Delta U = 0$.

It follows from the first law that

$$0 = Q - W \quad \text{that is} \quad Q = W.$$

So the effect of the cycle is that a net quantity, Q, of heat enters the gas, and an equal net quantity of work is done by the gas.

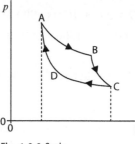

Fig. 1.8.9 Cycle

- The net work done by the gas over the cycle ABCDA is

$$W = \text{work done } by \text{ gas over ABC} - \text{work done } on \text{ gas over CDA}$$

$$= \text{area under ABC} - \text{area under ADC}$$

So $W = $ area inside 'loop' ABCDA

Example

Calculate the net heat taken in over the 'square' cycle shown.

Answer

net heat in = net work out

= area inside loop

$= 2.0 \times 10^{-3} \times 200 \times 10^3 \, \text{J} = 400 \, \text{J}$

(400 more joules of heat are taken in altogether along AB and DA than are given out altogether along BC and CD.)

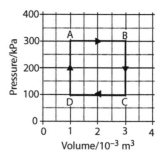

Fig. 1.8.10 Square cycle

1.8.6 Solids and liquids

Solids and liquids change their volumes very little (compared with gases), and therefore the work, W, that they do against external pressures is usually negligible, so $\Delta U = Q - W$ becomes simply

$$\Delta U = Q.$$

The heat input needed to raise the temperature of mass m of a solid or liquid by $\Delta\theta$ (provided this is not too large) is given (see Pointer) by

$$Q = mc\Delta\theta$$

in which c is a constant for the particular substance, called its **specific heat capacity**. UNIT: $\text{J kg}^{-1} \, °\text{C}^{-1} = \text{J kg}^{-1} \text{K}^{-1}$

Example

1.2 kg of tap-water at 15 °C is heated in a 3.0 kW electric kettle for 60 seconds. The heating 'element' is a coil of wire surrounded with thermally conducting, electrically insulating material in a metal case.

(a) Discuss the energy transfers in terms of *work*, *heat* and *internal energy* from the time the kettle is switched on.

(b) Calculate a value for the final temperature of the water, explaining why this value is likely to be too high. (shc of water: 4180 J kg^{-1} °C^{-1}).

Answer

(a) The electrical energy input to the heating element is classed as *work* input to the wire. In time Δt, $W = -3000$ J s$^{-1} \times \Delta t$.

For the first few milliseconds after switching on, the wire is at roughly the same temperature as its surroundings, so $Q = 0$ and $\Delta U = -W$, that is the work input goes to raising the wire's internal energy.

Soon the wire is significantly hotter than its surroundings, and heat starts to be conducted away from it. After a very short time the wire is so hot that heat flows out at the same rate that electrical work is being done, so the wire has stopped acquiring internal energy, and we have a 'steady state' when, $Q_{wire} = W = -3000$ J s$^{-1} \times \Delta t$.

The heat output from the wire is the heat input to the water, so, in the steady state, for the water, $\Delta U = Q_{water} = +3000$ J s$^{-1} \times \Delta t$. This ignores heat going to the material of the kettle, and heat escaping to the air.

(b) Assuming that all the energy input goes as heat to the water,

$$\Delta \theta = \frac{Q}{mc} = \frac{3000 \text{ W} \times 60 \text{ s}}{1.2 \text{ kg} \times 4180 \text{ J kg}^{-1} \text{ °C}^{-1}} = 36 \text{ °C}$$

Therefore, final temperature of water = 15 °C + 36 °C = 51 °C

In fact, the final temperature will be a little less mainly because:

- Some energy goes to raising the temperature of the kettle's heating element and the kettle itself.
- Some heat escapes from walls of the kettle into the surrounding air.

≫ **Pointer**

We usually use the Celsius temperature, θ, when dealing with solids and liquids. In fact, the K and the °C are the same size, so $\Delta \theta = \Delta T$ (even though θ/°C = T/K − 273.2). We could therefore also write $Q = mc\Delta T$.

≫ **Pointer**

The *Answer* uses $\Delta U = Q - W$ with sign conventions for Q and W given in Section 1.8.4.

quickꟼɪre

⑨ 3.0 kg of water (shc 4200 J kg^{-1} °C^{-1}) at 80.0 °C is poured into a saucepan of mass 1.5 kg made of stainless steel (shc 500 J kg^{-1} °C^{-1}) initially at 20.0 °C.

Calculate the new temperature.

>> *Pointer*

The water must be deep enough still to surround the trapped air when it has expanded by a third of its original length.

quicKpire

⑩ What is the point of stopping heating and waiting for the temperature to stabilise, before taking a pair of readings?

quicKpire

⑪ Express a from $T = al$ in terms of the cross-sectional area A of the trapped air, the amount n, the pressure p (and R).

quicKpire

⑫ Explain why, even if your value of θ_0 is close to −273°C, the uncertainty will be large.

1.8.7 Specified practical work

(a) Estimation of absolute zero of temperature using the gas laws

Fig. 1.8.11 shows an easy way to investigate the thermal expansion of a sample of air at constant pressure.

As it becomes hotter the trapped air pushes the sulphuric acid bead higher up the tube, keeping the air pressure very slightly above atmospheric. The concentrated sulphuric acid absorbs water vapour from the trapped air. **Wear safety spectacles** (although the acid should never escape).

Read temperature (θ) and length (l) of trapped air before heating, having stirred in (and melted) ice cubes if available. Heat for 30 seconds or so and stir until the temperature stabilises before starting to go down. Take another pair or readings. Carry on heating and taking readings aiming to go up in steps of about 10° each time, up to about 80°C.

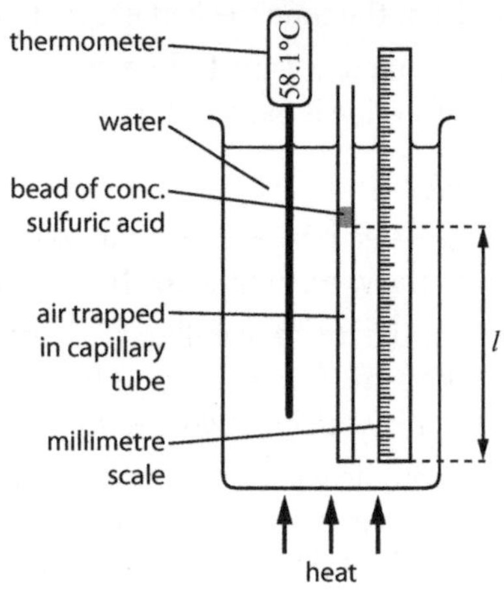

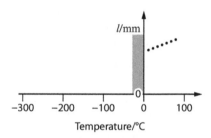

Fig. 1.8.11 Expansion of air at constant pressure

Fig. 1.8.12 Graph scales for expansion of air

We plot l / mm against θ /°C using scales as in Fig. 1.8.12. We draw the best-fit straight line and extrapolate (extend!) it back until it hits the temperature axis (at θ_0).

θ_0 is our value (in Celsius) for the absolute zero of temperature, when the air would occupy no volume.

We're assuming that the trapped air is behaving ideally, so that the kelvin temperature, defined to be proportional to pV for an ideal gas, is proportional to the air volume at constant pressure. In that case, $T = al$, in which a is a constant for the trapped air. The thermometer will have been calibrated in Celsius, defined by $\theta/°C = T/K - 273.2$. If $l = 0$, then $T = 0$, so $\theta = -273.2°C$. This, then, is the theoretical value of θ_0.

(b) Measuring the specific heat capacity of a solid

In Fig. 1.8.13 the thermometer and the heating coil are sunk in holes drilled in a block of the metal under investigation.

The mass, m, of the block needs to be known (using chemical balance or manufacturer's data).

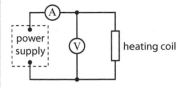

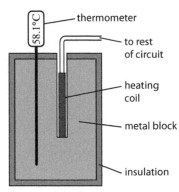

Fig. 1.8.13 Apparatus for finding the shc of a metal

The initial temperature (θ_0) of the block is read, then the power turned on and the current (I) and pd (V) read. After a measured time (t), the power is turned off and we wait for the temperature to stop rising. We read its maximum value (θ_1).

Assuming all the output from the heater goes as heat into the block,

$$VIt = mc(\theta_1 - \theta_0)$$

Hence c, the shc of the metal. See Section 1.8.6.

Graphical method

By taking pairs of readings of time t and temperature, θ, as θ rises from room temperature to (say) 90°C, we could obtain a value of c from the gradient of a graph of θ against t, having measured V, I and m.

This method may well give a less accurate value of c than the 'single-shot' method first described; the block will not be at a uniform temperature as readings are taken.

quickfire

⑬ Calculate the resistance of a heating coil which will deliver 50 W from a 12 V supply.

quickfire

⑭ Estimate how long it will take for a temperature rise of 30°C using a 50 W heater. (Typical mass of block: 1 kg; typical shc of metal: 400 J kg^{-1} °C^{-1}).

quickfire

⑮ For a one-off (non-graphical) determination of c, give an advantage and a disadvantage of using a rise of 70°C rather than of 20°C.

quickfire

⑯ Give an expression, in terms of m, c, V, and I, for the gradient of the graph of θ against t.

1. A sample of gas, initially at atmospheric pressure is contained in a metal cylinder with a leak-proof piston (Fig 1.8.14).
   (a) State how *in practice* you could increase the internal energy of the gas by:
      (i) doing work, with negligible heat flow
      (ii) heating, without doing work.
   (b) For each *method* in (a), state which of the quantities must be positive and which, if any, must be negative in the equation
      $$\Delta U = Q - W$$
      (i) doing work (with negligible heat flow)
      (ii) heating (without doing work).

Fig. 1.8.14 Gas sample

2. Some helium gas is contained in a cylinder fitted with a leak-proof piston. The piston is pushed in, causing the change **AB** shown in Fig. 1.8.15.

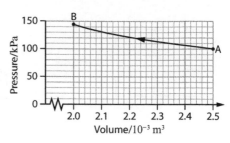

(a) At point **A** the gas temperature is 300 K. Calculate the amount of gas in moles.

(b) Calculate the gas temperature at **B**.

(c) Helium gas is monatomic. Calculate the change in its internal energy over **AB**.

(d) Determine directly from the graph a value for the work done on the gas over **AB**.

Fig. 1.8.15 Compression of a gas

(e) What does the first law of thermodynamics tell us about the heat flow over **AB**, and what can be deduced about how fast the piston is pushed in?

3. 0.040 mol of monatomic gas is taken through the cycle of changes shown in Fig. 1.8.16.

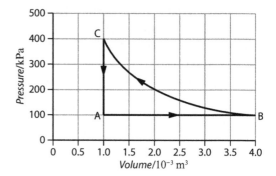

(a) Determine the temperatures at **A**, **B** and **C**.

(b) Complete the table using quantities ΔU, W and Q as defined for use in the *first law of thermodynamics*.

	AB	BC	CA	ABCA
$\Delta U / J$				
W / J		−555		
Q / J				

Fig. 1.8.16 Cycle for a gas

4. A student performs the experiment described in Section 1.8.7 (a). Her results are listed below, together with her estimates of their uncertainties.

$\theta / °C$	±0.1 °C	19.2	31.4	42.8	50.5	60.8	69.1	81.3
l / mm	±0.5 mm	48.0	50.0	52.0	53.0	55.0	56.5	58.5

Determine the value that these results suggest for the absolute zero of temperature in °C, and its absolute uncertainty, as follows:

(a) Choosing suitable scales, plot the values of l against θ. Include error bars. [There is no need to include zero for either l or θ on your scales.]

(b) Determine the equation of both the steepest and least steep graphs consistent with the error bars, in the form $l = m + c$.

(c) Hence determine the equation of the best-fit graph in the form
$l = (m \pm \Delta m) + (c \pm \Delta c)$.

(d) Use your equations for the steepest, least steep and best fit graphs to obtain maximum, minimum and best values for the absolute zero of temperature in °C, together with the uncertainty in the best value.

Component 1 Newtonian physics Summary

1.5 Circular motion

- The radian measure of angle
- Angular velocity, period, frequency
$$\omega = 2\pi f = \frac{2\pi}{T}$$
- The relationship between speed and angular speed $v = r\omega$
- The centripetal acceleration is the acceleration
- towards the centre and is given by $a = r\omega^2 = \frac{v^2}{r}$
- The centripetal force is the resultant force acting on a body moving at a constant speed in a circle;
$$F = mr\omega^2 = \frac{mv^2}{r}$$

1.6 Vibrations

- Definition of simple harmonic motion (shm): $a = -\omega^2 x$
- Interpretation and use of $x = A\cos(\omega t + \varepsilon)$ and $v = -A\omega \sin(\omega t + \varepsilon)$
- The terms frequency, amplitude and phase
- Interpretation and use of $T = 2\pi\sqrt{\dfrac{m}{k}}$ and
$$T = 2\pi\sqrt{\frac{l}{g}}$$
- KE and PE interchange in SHM: $E_k - t$, $E_p - t$, $E_k - x$ and $E_p - x$ graphs
- Free oscillation, damped oscillations; practical examples
- Forced oscillations with damping; Graph of amplitude against driving frequency
- Resonance

1.7 Kinetic theory of gases

- Equation of state for an ideal gas as $pV = nRT$ and $pV = NkT$
- The kelvin temperature scale
- Assumptions of the kinetic theory of gases
- Pressure caused by molecular motion:
$$p = \tfrac{1}{3}\rho\overline{c^2} = \tfrac{1}{3}\frac{N}{V}m\overline{c^2}$$
- The mole, molar mass, M, the relative molar mass, M_r, the Avogadro constant, N_A, the molar gas constant, R, and the Boltzmann constant, k
- Mean translational KE per molecule $= \tfrac{3}{2}kT$ and its derivation

1.8 Thermal physics

- Thermodynamic systems; the internal energy, U, of a system
- Systems have minimum energy at absolute zero
- For an ideal monatomic gas, $U = \tfrac{3}{2}nRT$
- Thermal equilibrium; heat flow, Q, produced by temperature difference
- Energy transfer by work; $W = p\Delta V$ for the work done by a gas
- The first law of thermodynamics, $\Delta U = Q - W$, where Q is the heat input to the system and W the work done by the system
- For solids and liquids, W is usually negligible so $Q = \Delta U$
- Specific heat capacity for a solid or liquid; $Q = mc\Delta\theta$

Specified practical work

- Measurement of g with a pendulum
- Investigation of the damping of a spring
- Estimation of absolute zero by the use of the gas laws
- Measurement of the specific heat capacity of a solid

Component 2 Knowledge and Understanding

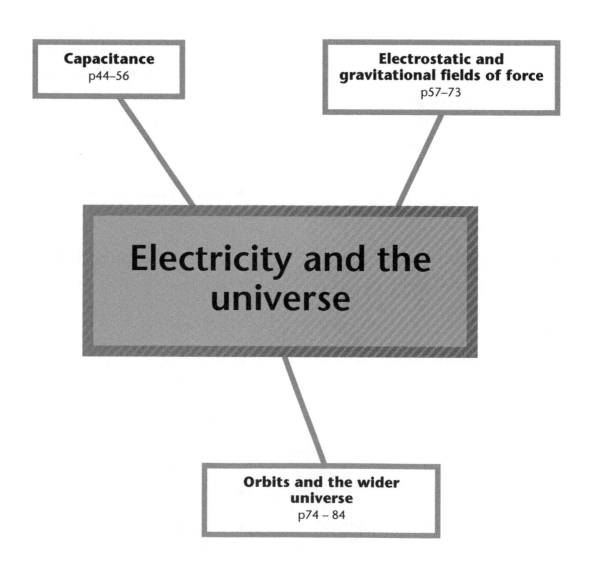

Capacitance
p44–56

Electrostatic and gravitational fields of force
p57–73

Electricity and the universe

Orbits and the wider universe
p74 – 84

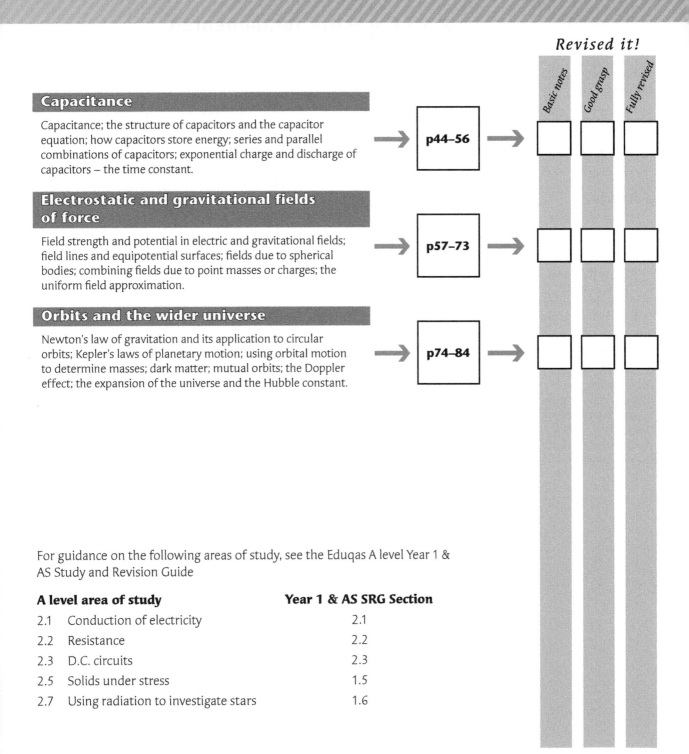

Revised it!

Basic notes   Good grasp   Fully revised

Capacitance

Capacitance; the structure of capacitors and the capacitor equation; how capacitors store energy; series and parallel combinations of capacitors; exponential charge and discharge of capacitors – the time constant.

p44–56

Electrostatic and gravitational fields of force

Field strength and potential in electric and gravitational fields; field lines and equipotential surfaces; fields due to spherical bodies; combining fields due to point masses or charges; the uniform field approximation.

p57–73

Orbits and the wider universe

Newton's law of gravitation and its application to circular orbits; Kepler's laws of planetary motion; using orbital motion to determine masses; dark matter; mutual orbits; the Doppler effect; the expansion of the universe and the Hubble constant.

p74–84

For guidance on the following areas of study, see the Eduqas A level Year 1 & AS Study and Revision Guide

2.4 Capacitance

2.4.1 Capacitor fundamentals

Capacitors are simple devices that store charge. They are made of two parallel metal plates separated by an insulator. For A-level calculations, you will usually have air or a vacuum between the plates. In practice, capacitors will have other insulators (called dielectrics) between the plates. Dielectrics increase capacitance and some can increase capacitance by a factor of thousands.

Key Term

Capacitance =

$$\frac{\text{charge on either plate}}{\text{pd between plates}}$$

UNIT: F (farad) [= C V^{-1}]

≫ Pointer

You can also define capacitance by giving the equation $C = \frac{Q}{V}$ and defining the terms. i.e.
C = capacitance
Q = charge on either plate
V = pd across plates.

Key Term

Dielectric = insulator between the plates of a capacitor, also serving to make the capacitance larger than if there were just empty space between the plates.

Fig. 2.4.1 Capacitor structure

When a pd is applied to a capacitor, charge is transferred by the power supply around the circuit. The plates then carry an equal and opposite charge (the net charge being zero in agreement with conservation of charge from AS SRG Section 2.3).

Fig. 2.4.2 Charges on capacitor plates

The quantity of charge Q on each plate depends on the pd applied and the size of the capacitor. In fact it's proportional to both of them.

$$Q = CV$$

In the Data booklet you'll see $C = \frac{Q}{V}$ and this is used as a definition of capacitance.

There is a misconception that a capacitor is similar to a cell. This is not true. This misconception arises from poor use of terms regarding a 'charged' and 'discharged' cell – the person who decided to call cells or batteries 'charged' was a non-physicist! You should consider a cell to be a 'pump' that can provide a flow of charge until it has run out of chemical energy. A capacitor, on the other hand, really does require charging using a pd.

2.4.2 Capacitance

This is the equation that you'll need to use in order to calculate the capacitance of actual metal plate capacitors:

$$C = \frac{\varepsilon_0 A}{d}$$

where A is the area of the plates, d is the distance between the plates and ε_0 is the permittivity of free space that you will also encounter in Section 2.6. Three things to note here:

1 Capacitance is proportional to the plate area.

2 Capacitance is inversely proportional to separation of the plates.

3 You will only ever need to do calculations based on capacitors with air or vacuum between the plates so that you can use ε_0 (8.85×10^{-12} F m^{-1}) which you'll see on the front of the Data booklet).

Example – a typical easy starter question on capacitors:

A capacitor consists of two square metal plates of side 12.0 cm, separated by 0.078 mm.

Calculate:

(i) the capacitance of the capacitor

(ii) the charge stored on the capacitor when charged by a 14.3 V cell.

Answer

(i) $C = \dfrac{\varepsilon_0 A}{d} = \dfrac{8.85 \times 10^{-12} \times 0.12 \times 0.12}{0.078 \times 10^{-3}} = 1.63 \times 10^{-9}$ F (or 1.63 nF)

(ii) $Q = CV = 1.63 \times 10^{-9} \times 14.3 = 2.33 \times 10^{-8}$ C (or 23.3 nC)

The most difficult thing about these types of question is getting the unit conversions and the powers of 10 correct.

2.4.3 Energy of a capacitor

The energy stored by a capacitor (or its internal energy U) is given by the equation:

$$U = \tfrac{1}{2}QV$$

This is the equation that appears in the Data booklet but it can be combined with $Q = CV$ to provide another two equations:

$$U = \tfrac{1}{2}QV = \tfrac{1}{2}CV^2 = \tfrac{1}{2}\frac{Q^2}{C}$$

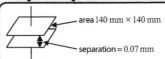

quickfire

area 140 mm × 140 mm

separation = 0.07 mm

① Calculate:
   i) the capacitance
   ii) the pd required to store a charge of 211 nC.

quickfire

area 2.0 cm × 2.0 cm

separation = 0.12 mm

② Calculate the charge held by the capacitor when charged by a cell of emf 1.6 V.

quickfire

area 3.8 cm × 3.8 cm

③ The capacitor is charged by a 12 V power supply and stores a charge of 84 nC. Calculate the separation of the plates.

quickfire

④ A 470 μF capacitor is charged using a cell of emf 9.52 V and is then discharged through a resistor of 0.24 Ω. Calculate:

i)   the charge stored by the charged capacitor

ii) the energy stored by the charged capacitor

iii) the initial current when the capacitor starts to discharge.

≫ **Pointer**

The concept of the electric field is developed in Section 2.6. As far as this topic is concerned we can regard it as the potential gradient, i.e. the pd per unit distance: Hence the unit, $V\,m^{-1}$.

Here's a quick way of demonstrating the charge and energy that a capacitor can store. First, charge a large capacitor (~1000 μF) using a 9 V cell. Then disconnect the cell. Note that the charge held by the capacitor must remain after it has been disconnected because it has nowhere to go. Then touch the two legs of the capacitor together to discharge it (short circuit). You should see a tiny little spark due to the large discharge current. How much charge and energy was stored by the 1000 μF capacitor?

$$Q = CV = 1000 \times 10^{-6} \times 9 = 0.0090\,C \quad \text{or } 9.0\,mC$$

and
$$U = \tfrac{1}{2}CV^2$$

$$U = \tfrac{1}{2} \times 1000 \times 10^{-6} \times 9^2 = 0.0405\,J$$

These are not very large numbers even though 1000 μF (1 mF) is a large capacitor for electronics. However, the current is a different matter altogether. When the legs of the capacitor are 'shorted' there is no resistance in the circuit – not even an internal resistance of a cell (see AS SRG Section 2.3). The only resistance present is the resistance of a few centimetres of tinned wire coming from the capacitor which will be ~0.01 Ω. With an initial pd of 9 V, this gives an initial current of

$$I = \frac{V}{R} = \frac{9}{0.01} = 900\,A.$$

2.4.4 Electric field between the plates of a capacitor

Another tricky concept is the electric field, or E-field (see **Pointer**). The E-field is uniform between capacitor plates and is given by the equation $E = \dfrac{V}{d}$.

— area 3.8 cm × 3.8 cm

Fig. 2.4.3 Electric field between capacitor plates

There aren't many questions that an examiner can ask regarding the electric field in a capacitor but here's one that will set off some more sparks. It deals with the fact that, when air is subjected to an electric field greater than $3 \times 10^6\,Vm^{-1}$ the air itself is said to 'break down' and a spark will jump across the gap of the field (this is what also happens in lightning).

Example

A parallel plate capacitor has square-shaped plates with air between them. The pd is increased across its plates until a spark discharge occurs when the pd is 540 V. The charge stored by the capacitor just before the spark appears is 8.7 μC. What are the dimensions of the capacitor?

quickfire

⑤ Calculate the separation of the plates of a capacitor given that a spark jumps across the plates when $V = 1650\,V$.

(The 'breakdown' value of E in air = $3 \times 10^6\,V\,m^{-1}$)

Answer

First, find the separation of the plates using the equation for the field:

$$E = \frac{V}{d} \quad \rightarrow \quad d = \frac{V}{E} = \frac{540\,\text{V}}{3 \times 10^6\,\text{V m}^{-1}} = 1.8 \times 10^{-4}\,\text{m} \ (0.18\,\text{mm})$$

You can also calculate the capacitance:

$$C = \frac{Q}{V} = \frac{8.7 \times 10^{-6}\,\text{C}}{540\,\text{V}} = 1.61 \times 10^{-8}\,\text{F} \ (16.1\,\text{nF})$$

Now you can calculate the area of the plates:

$$C = \frac{\varepsilon_0 A}{d} \quad \rightarrow \quad A = \frac{Cd}{\varepsilon_0} = \frac{1.61 \times 10^{-8}\,\text{F} \times 1.8 \times 10^{-4}\,\text{m}}{8.85 \times 10^{-12}\,\text{F m}^{-1}} = 0.327\,\text{m}^2$$

So, because the question stated that the plates were square
length of sides of plates = $\sqrt{0.327} = 0.572$ m (or 57.2 cm)
So the dimensions of the capacitor (in mm) are $572 \times 572 \times 0.18$.

2.4.5 Combinations of capacitors

As with resistors, capacitors can be combined either in parallel or in series. Although the formulae for combining capacitors are the same, **the equations for the parallel and series cases are swapped**.

(a) Capacitors in series

For an arrangement of three **series** capacitors as shown the overall capacitance, C, is given by

$$\frac{1}{C} = \frac{1}{C_1} + \frac{1}{C_2} + \frac{1}{C_3}$$

which is of the same form as the equation for resistors in parallel. This means that the overall capacitance is always less than the smallest capacitor, i.e. in a series combination you can't store as much charge – you can think of it as 'increasing' the separation of the plates and decreasing the overall capacitance.

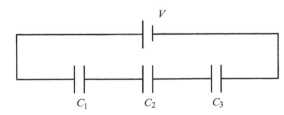

Fig. 2.4.4 Capacitors in series

quickfire

⑥ The energy stored in the capacitor in Quickfire 5 just before sparking is 4.2 J. Calculate the capacitance and the area of the plates. Is the separation and area of these plates achievable in your lab using tin foil?

 Grade boost

For capacitors in series, the pd is shared between them exactly as the pd is shared between resistors in series. However, the pd is shared in inverse proportion to the capacitance (because if you have half the capacitance you need twice the pd to hold the same charge).

 Pointer

Due to conservation of charge each capacitor in **series** must hold **exactly the same charge**.

 Grade boost

As with resistors in parallel, a good equation to remember for two capacitors in series is:

$$C = \frac{C_1 C_2}{C_1 + C_2}.$$

Warning: it only works for **two** capacitors.

≫ Pointer

For capacitors in parallel, the pd across each of the parallel branches is exactly the same – similar to resistors in parallel. However, the charge stored is proportional to the capacitance of the parallel branch (because $Q = CV$ and V is a constant for each branch).

Grade boost

When combining capacitors (in series and parallel) there's no need to convert the µF or nF to F (as long as they are all in the same units). The answer will come out in these same units.

quicKfire

⑦ Calculate the overall capacitance

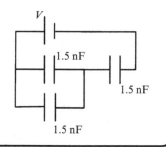

≫ Pointer

Alternative working in Example (ii)

$$C = \frac{C_1 C_2}{C_1 + C_2} = \frac{6.2 \times 4.7}{6.2 + 4.7}$$

$$= 2.7 \text{ nF (try it!)}$$

(b) Capacitors in parallel

For an arrangement of three **parallel** capacitors as shown the overall capacitance, C, is given by $C = C_1 + C_2 + C_3$

This is just the same as for resistors in series, you simply add the capacitances to obtain the overall capacitance. A good way of looking at this is that your parallel capacitors are, effectively, one big capacitor with a bigger area. Hence, the capacitance increases and you can store more charge.

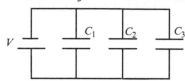

Fig. 2.4.5 Capacitors in parallel

Example

Calculate the overall capacitance of the following two combinations:

(i)   (ii)

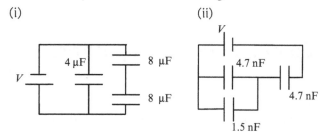

Fig. 2.4.6 Series and parallel capacitor combinations

Answer

(i)   The 8 µF in series with the other 8 µF gives 4 µF (no need for a calculator you should see that it's the same as two equal resistors in parallel). You now have two 4 µF capacitors in parallel which gives 8 µF (it would have been far more sensible just to have the one 8 µF capacitor in the first place).

(ii) First, the 4.7 nF in parallel with the 1.5 nF is easy, you add them and get an overall capacitance of 6.2 nF. Then you need to calculate the overall capacitance of 6.2 nF in series with 4.7 nF.

$$\frac{1}{C} = \frac{1}{C_1} + \frac{1}{C_2} = \frac{1}{6.2} + \frac{1}{4.7} = 0.374 \quad \rightarrow \quad C = \frac{1}{0.374} = 2.7 \text{ nF}$$

(see **Pointer**).

2.4.6 Discharging a capacitor through a resistor

(a) How the circuit works

The standard circuit for charging and discharging a capacitor is shown in Fig. 2.4.7. When the switch is up, the capacitor charges. When the switch is down, the capacitor discharges through the resistor. The charged capacitor provides a pd across the resistor which gives a current.

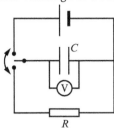

Fig. 2.4.7 Circuit for discharging a capacitor

The current through the resistor is $\dfrac{\Delta Q}{\Delta t}$ and, from the point of view of the capacitor, this is the rate at which the capacitor is **losing** charge. For the capacitor you can write:

$$\frac{\Delta Q}{\Delta t} = -\,\text{current} = -\frac{V}{R} = -\frac{Q}{RC} \qquad \text{because } V = \frac{Q}{C}$$

The equation $\dfrac{\Delta Q}{\Delta t} = -\dfrac{Q}{RC}$ tells you that the capacitor is losing charge at a rate that is proportional to the charge on the capacitor $\left(\dfrac{\Delta Q}{\Delta t}\right) \propto -Q$. So, when the capacitor is fully charged, it loses charge quickly. As the charge decreases, the capacitor then loses charge at a slower rate.

If you plot a graph of charge on the capacitor against time, the charge must go from a certain value (Q_0 say) to zero. But you also know that the gradient of the line $\left(\dfrac{\Delta Q}{\Delta t}\right)$ is always decreasing (because $\dfrac{\Delta Q}{\Delta t} \propto -Q$.). You can probably guess the shape of the graph – it's an exponential decay. This is always true when you have a variable that is decreasing at a rate proportional to the variable itself. In this case, the charge is decreasing at a rate proportional to the amount of charge held. A similar case (to a reasonable approximation) is the flow of water out of a burette – the rate of decrease of height of water is proportional to the height of the water. Similarly, in radioactive decay the rate of decrease of radioactive nuclei is proportional to the number of radioactive nuclei present. All these examples give exponential decays.

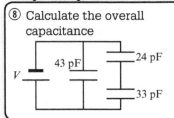
Grade boost

Remember the circuit for charging and discharging a capacitor, it's useful: it turns up both in theory papers and in your practical work.

» Pointer

If you're doing A-level Mathematics you should be able to integrate this equation but from a physics point of view it's better to understand what the equation means and why the final answer makes sense.

quickpire

⑨ A discharging capacitor loses charge at a rate of 15 mA (i.e. 15 mC s^{-1}) when it stores a charge of 1.0 C. What will its rate of discharge be (through the same resistor) when it stores (a) 0.5 C, (b) 1.0 mC?

Grade boost

Get used to using the exponential equation it comes up many times in A2 (discharging, radioactivity, biological measurement and in practical work). You'll also have to be able to take logs of the exponential equation to calculate times.

» *Pointer*

Or: $\dfrac{18\,\mu C}{12\,V} = 1.5\,\mu F$

There's no need to convert to F here.

Key Term

Time-constant for a discharging or discharging capacitor = RC

In one time-constant, the capacitor loses 63% of its charge, i.e. there's 37% left.

quickfire

⑩ Do part (ii) of the example by calculating 37% of 18 μC then reading the time-constant (RC) off the graph. Equate this value to RC to find a value for R.

(b) Discharge equation and graph

The equation for the discharging capacitor is $Q = Q_0 e^{-\frac{t}{RC}}$ and the graph for one particular RC combination is shown in Fig. 2.4.8.

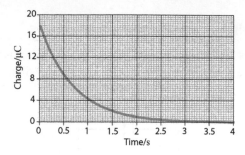

Fig. 2.4.8 Capacitor discharge curve

Example

In the discharge graph given, the pd used to charge the capacitor was 12.0 V. Use values from the graph to calculate the value of:

(i) the capacitance, C

(ii) the resistance, R

(iii) the time-constant, RC.

Answers

(i) From the graph the initial charge on the capacitor is 18 μC

$$\text{Then, } C = \frac{Q}{V} = \frac{18 \times 10^{-6}\,C}{12\,V} = 1.5 \times 10^{-6}\,F$$

(ii) First rearrange a little: $Q = Q_0 e^{-\frac{t}{RC}} \rightarrow \dfrac{Q}{Q_0} = e^{-\frac{t}{RC}}$

now take logs $\quad \ln \dfrac{Q}{Q_0} = \ln e^{-\frac{t}{RC}} = -\dfrac{t}{RC}$

Now all you need to do is plug in some numbers:

$Q_0 = 18$ μC, then you need Q and t from the graph. There's no obvious choice but $t = 1.0$ s and $Q = 4.4$ μC give:

$$R = \frac{-1}{1.5 \times 10^{-6} \times \ln(4.4/18)} = 4.73 \times 10^5\,\Omega \text{ (or 470 k}\Omega)$$

That was a lot of algebra just to find the resistance but don't worry, this type of manipulation is the most difficult that you'll ever come across at A-level.

(iii) This is a lot easier:

Time-constant = $RC = 4.73 \times 10^5\,\Omega \times 1.5 \times 10^{-6}\,F = 0.71$ s

You can do a quick check that the time-constant is correct because after 1 time-constant the capacitor should be 63% discharged (this is something that all electronics people memorise).

Look at the graph carefully: the charge is around 6.5 μC after 0.71 s. As a percentage of 18 μC, this is $\dfrac{6.5}{18} \times 100 = 36\%$

i.e. the capacitor is $(100 - 36) = 64\%$ discharged which is close enough to 63% considering the size of the graph. Had you known this fact about the time-constant before doing part (ii), you could have saved a lot of time and effort (see **Quickfire 10**).

2.4.7 Charging a capacitor through a resistor

A circuit for charging a capacitor through a resistor is shown in Fig. 2.4.9. The capacitor starts charging as soon as the switch is closed.

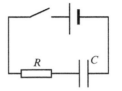

Fig. 2.4.9 Charging circuit

(a) The charging curve

Initially, the current is large because all the emf of the cell is across the resistor – the capacitor initially holds no charge and hence its pd is zero. As the charge on the plates of the capacitor increases (due to the current), its pd increases and the pd across the resistor decreases. Since the pd across the resistor is decreasing the current is also decreasing. The current then continues to decrease until the capacitor is fully charged, all the pd is across the capacitor and the current has dropped to zero (Fig. 2.4.10).

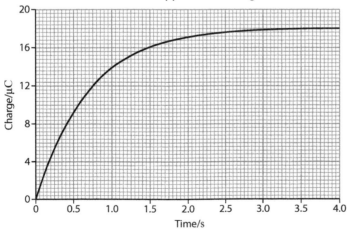

Fig. 2.4.10 Capacitor charging curve

⑪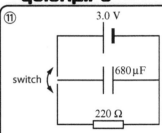

i) In the circuit, how do you charge then discharge the capacitor?
ii) Calculate the time-constant for discharging.
iii) Calculate the initial charge on the capacitor.
iv) Calculate the time the capacitor takes to lose half its charge.
v) What percentage of the energy of the capacitor remains when it has lost half its charge?

quickfire

⑫ Explain why the capacitor in Quickfire 11 charges very quickly (in your answer you should refer to the time-constant).

Grade boost

Remember that $e^0 = 1$ so that $e^{-\frac{t}{RC}}$ is always 1 when $t = 0$. This also means that $1 - e^{-\frac{t}{RC}}$ is always 0 when $t = 0$.

Grade boost

Remember that $e_0^{-\infty} = 0$ so that $e^{-\frac{t}{RC}}$ is 0 when a long time has passed and so $Q = Q_0$.

quickfire

⑬ An initially uncharged 1.5 nF capacitor is charged through a 28 kΩ resistor using a battery of emf 9.5 V. Calculate:
a) the time-constant
b) the pd across the capacitor after 1 time-constant
c) the pd across the capacitor after 90 μs
d) the time it takes for the capacitor to store 32 nJ.

quickfire

⑭ In Fig. 2.4.11, write down the equation for the variation of the current in the resistor.

The equation that gives the time variation of the charge on the charging capacitor is:

$$Q = Q_0\left(1 - e^{-\frac{t}{RC}}\right)$$

and the shape of the graph is the same as for discharging except that it is upside down! The time-constant is still given by RC but this is the time for the capacitor to become 63% **charged** (instead of discharged).

(b) The variation of charging pd and current with time

If you divide the above equation by the capacitance of the capacitor you get:

$$\frac{Q}{C} = \frac{Q_0}{C}\left(1 - e^{-\frac{t}{RC}}\right)$$

With a bit of luck, you'll remember that $V = \frac{Q}{C}$, which gives us:

$$V = V_0\left(1 - e^{-\frac{t}{RC}}\right)$$

This means that the pd across the capacitor follows exactly the same shape as the charge with exactly the same time-constant. Note that V_0 will be the emf of the cell.

However, the current in the circuit does not follow the same pattern. The current starts as a maximum and drops to zero. Anyone doing A-level Maths should be able to obtain $\frac{dQ}{dt}$ from the Q equation to give:

$$I = I_0 e^{-\frac{t}{RC}}$$

Perhaps surprisingly, the current follows the same pattern whether the capacitor is charging or discharging. Again I_0 will be given by $\frac{V_0}{R}$ which can also be written as $\frac{Q_0}{RC}$

2.4.8 Specified practical work

(a) Investigation of the charging and discharging of a capacitor to determine the time-constant

This experiment is perhaps most easily done for discharging a capacitor using the circuit shown in Fig. 2.4.11. In order to take readings at a sensible rate, a time-constant RC of at least 30 s is required. This means that the resistor must be around 150 kΩ or more and the capacitor around 220 μF or more.

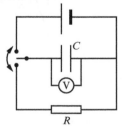

Fig. 2.4.11 Determining the time-constant

Procedure:

- Set up the apparatus as shown in the circuit diagram, **ensuring that the capacitor is connected +ve to +ve** (otherwise it might explode).
- Push the switch up to charge the capacitor.
- Push the switch down and start the stopwatch at the same time.
- Record pd values from the voltmeter at regular intervals (e.g. 10 s – 30 s).
- Continue until around three time-constants have passed (12.5% charge remaining).
- Repeat experiment again to estimate random uncertainties, obtain mean values, check for anomalies, etc.

Theory

While discharging, the pd across the capacitor will be given by $V = V_0 e^{-\frac{t}{RC}}$

Taking natural logs gives

$$\ln V = \ln V_0 - \frac{t}{RC}$$

Hence a graph of $\ln V$ against t should be linear with a gradient of $-\dfrac{1}{RC}$ and an intercept of $\ln V_0$ on the $\ln V$ axis Fig. 2.4.12 is a typical graph. The time-constant is thus minus the reciprocal of the gradient.

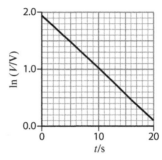

Fig. 2.4.12 Graph of $\ln V$ against t

Alternative circuits

The circuits in Fig. 2.4.13 can also be used to determine the time-constant for a discharging capacitor. Look at them carefully and make sure you know how you could use them.

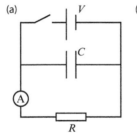
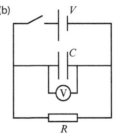

Fig. 2.4.13 Alternative circuits for capacitor discharge

Charging circuits

The following circuits can be used to investigate the charging of a capacitor:

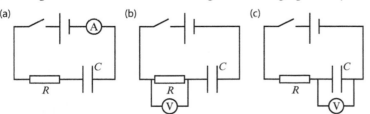

Fig. 2.4.14 Alternative circuits for investigating charging a capacitor

MATHS TIP

For using logs and to linearise exponential relationships see Section M.3

quickfire

⑮ The resistor in the discharging circuit has a value of $1.5\,k\Omega$. The graph of $\ln V$ against t is shown in Fig. 2.4.12. Calculate:
a) the emf of the power supply
b) the time-constant
c) the capacitance.

quickfire

⑯ Write the relationship between I and t for a discharging capacitor.

For Fig. 2.4.13 (a) when is $I = 0$ in the equation you have written?

quickfire

⑰ What are the values of V_0 and I_0 in the circuits in Figs. 2.4.13 (b) and (a) respectively?

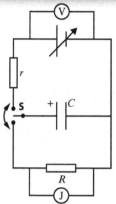

Fig. 2.4.15 Investigating the energy stored in a capacitor

Circuit (a) produces the familiar relationship $I = I_0 e^{-\frac{t}{RC}}$, though we cannot actually measure the initial current, I_0, because of the ammeter's response time. The analysis is as for the discharging circuit.

For circuit (c), the relationship is $V = V_0\left(1 - e^{-\frac{t}{RC}}\right)$ as we have seen. The only way of finding the time-constant, RC, is to plot a graph of V against t and find the time for the pd to drop to 37% of its initial value.

(b) Investigation of the energy, U, stored in a capacitor

The simplest method is using a smoothed variable DC supply and an energy meter – a.k.a. joule meter (Fig. 2.4.15).

Method

1. Connect up the circuit as shown, taking care to observe the polarity of the capacitor, C.

2. Set the power supply to a suitable low pd (e.g. 2 V), and charge the capacitor through the resistor, r, by moving the switch, **S**, to the upper position.

3. Discharge the capacitor through the resistor, R, wait until the reading on the joule meter has settled down* and record this reading.

4. Repeat 2–3 times at this pd

5. Repeat steps 2–4 for a series of different pds, e.g. in 2 V steps up to 12 V, taking care not to exceed the rated pd of the capacitor.

* The time for the reading to settle and reach its final value will depend upon the time-constant, RC. A good rule of thumb is 10 time-constants (see Quickfire 19). The role of the charging resistor, r, is just to prevent too great a current surge when charging. Here, too, the 10 time-constants rule should be observed.

Analysis

We expect the relationship to be $U = \frac{1}{2}CV^2$ so a graph of U against V^2 should be a straight line through the origin with a gradient of $\frac{1}{2}C$.

1. A capacitor is made from two rectangular metal plates of dimensions 32.0 cm × 22.0 cm separated by a 0.22 mm air gap. Calculate:

   (a) (i) the capacitance

      (ii) the charge, Q, held by the capacitor when a pd of 44 V is placed across it

      (iii) the energy, U, stored by the capacitor when a pd of 44 V is placed across it

      (iv) the maximum charge that can be held by this capacitor, assuming a maximum electric field of 3×10^6 V m^{-1}.

   (b) The separation of the plates of the capacitor can be varied. Explain why the maximum charge that the capacitor can hold is independent of the separation (assuming a maximum electric field of 3×10^6 V m^{-1}).

2. Calculate the capacitances of the capacitor combinations shown

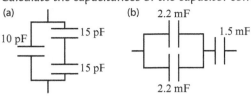

(a)

(b) 2.2 mF

(c) 2.2 mF   1.5 mF 1.5 mF

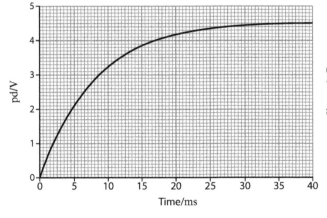

[Hint: Save time in part (c) by using your answers to (a) and (b)]

3. A 2.7 μF capacitor is charged then discharged using the circuit shown. The initial discharging current, I_0, is 50 μA. Calculate:

   (a) the charge held by the capacitor after the switch is pushed upwards

   (b) the resistance R

   (c) the time-constant of the discharging circuit

   (d) the charge held by the capacitor when it has been discharging for 300 ms

   (e) the time taken for the capacitor to lose 85% of its charge.

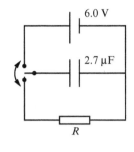

4. A capacitor C is charged through a resistor, R. The variation of charge stored and pd with time is shown in the two graphs.

   Use the graphs to calculate:

   (a) the pd of the supply V

   (b) the time-constant of the circuit

   (c) the resistance R (use a suitable tangent)

   (d) the capacitance C.

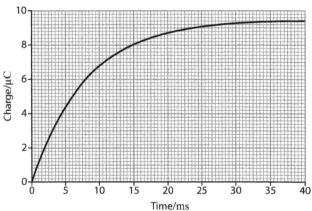

5. A capacitor is charged and discharged using the circuit shown.

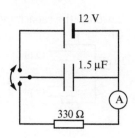

   (a) Explain why the time-constant for charging is (close to) zero.

   (b) Show that the time-constant for discharging is approximately half a microsecond.

   (c) Calculate the pd remaining on the capacitor after it has been discharging through the resistor for (i) 1 μs, (ii) 10 μs, (iii) 0.5 ms.

   (d) The switch can be forced to oscillate up and down by sound waves.

      (i) In light of your answer to part (c), explain why the mean current in the ammeter is $Q_0 f$, where f is the frequency of the sound waves.

      (ii) Calculate the mean current detected by the ammeter when sound waves of frequency 1 kHz are incident upon the switch.

   (e) This type of circuit is used as a frequency to current converter. Explain why the behaviour of this device starts becoming non-linear around a frequency of 1 MHz.

   (f) Sketch a graph of the pd across the capacitor when ultrasound waves of frequency 1 MHz are incident upon the switch.

6. A student investigates the energy stored in a capacitor as described in Section 2.4.8(b). He uses a capacitor which is labelled 47 mF with a tolerance of 20% and a 10 Ω discharge resistor (the tolerance of which may be ignored).

   (a) Estimate the time-constant of the discharge using the above information.

   (b) The gradient of the U v V^2 graph was (0.028 ± 0.002) J V^{-2}.
       Explain whether this is consistent with the above information.

   (c) The student applied the 10 time-constants rule based upon the component labels. Calculate the fraction of the energy recorded at this time and comment on how this will affect the student's calculated value of the capacitance.

2.6 Electrostatic and gravitational fields of force

These two fields are quite different in origin, but they have important features in common: the *inverse square law* and the applicability of the concepts of *field strength* and *potential*. Sections 2.6.1 – 2.6.5 deal with electric fields. Sections 2.6.6 – 2.6.10 deal with gravitational fields, but more briefly where the ideas are very similar to those met earlier. Section 2.6.11 is a useful summary table from the specification.

2.6.1 Electric field strength

We can test for the presence of an electric field using a positive charge, q. We shall call this a 'test charge'.

If q experiences a force proportional to its charge, q, we say it is in an electric field. For example, the test charge will reveal an electric field near a working Van de Graaff generator (see diagram).

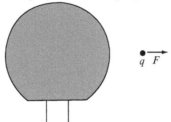

The **electric field strength** or **electric intensity**, E, at a point is defined in words under **Key terms**, or as an equation:

Fig. 2.6.1 Testing for an electric field

$$E = \frac{\text{force on (positive) test charge}}{\text{test charge}} \quad \text{that is} \quad E = \frac{F}{q}$$

- *units*: $N\,C^{-1}$ ($= V\,m^{-1}$) See **Pointer**.
- Electric field strength is a vector (because force is a vector).
- We define E as F/q because twice the test charge will feel twice the force, so F/q doesn't depend on q, but only on the field that q is in.
- We often need the equation re-arranged as...

$$F = qE$$

Example

A test charge of 5.0 nC experiences a force of 0.40 mN at a point near a Van de Graaff sphere. Calculate the electric field strength.

Answer

$$E = \frac{F}{q} = \frac{0.4 \times 10^{-3}\,N}{5.0 \times 10^{-9}\,C} = 80\ \text{kN C}^{-1} \quad \text{(that is 80 kV m}^{-1}\text{)}$$

» Pointer

You'll soon see the sense in using $V\,m^{-1}$ as a unit for E (Section 2.6.3 (d)), but, for now, note that
$$V\,m^{-1} = J\,C^{-1}\,m^{-1}$$
$$= N\,m\,C^{-1}\,m^{-1} = N\,C^{-1}$$

» Pointer

It's perfectly possible to use a negative test charge. E is then in the opposite direction to the force on the test charge.

quickfire

① Typically, there is a naturally arising downward electric field of strength $120\,V\,m^{-1}$, just above the Earth's surface. If there is a water droplet here, with a charge of -2.0 nC, find:

a) the electric force it experiences

b) the droplet's initial acceleration if its mass is 8.0×10^{-9} kg. (Don't forget the droplet's weight.)

Example

Calculate the acceleration of a positive ion of mass 4.65×10^{-26} kg and charge 3.20×10^{-19} C placed at the same point as in the previous example. (Assume forces other than from the electric field are negligible.)

Answer

Force on ion = mass × acceleration

So $\qquad qE = ma$

So $\qquad a = \dfrac{qE}{m} = \dfrac{3.20 \times 10^{-19}\,\text{C} \times 80 \times 10^{3}\,\text{N C}^{-1}}{4.65 \times 10^{-26}\,\text{kg}} = 5.5 \times 10^{11}\,\text{m s}^{-2}$

2.6.2 Coulomb's Law

Two 'point charges' (compact charges), Q_1 and Q_2, separated by a distance r, in a vacuum (or air) exert forces on each other as shown. Experimentally, we find:

$$F = \frac{1}{4\pi\varepsilon_0}\frac{Q_1 Q_2}{r^2}.$$

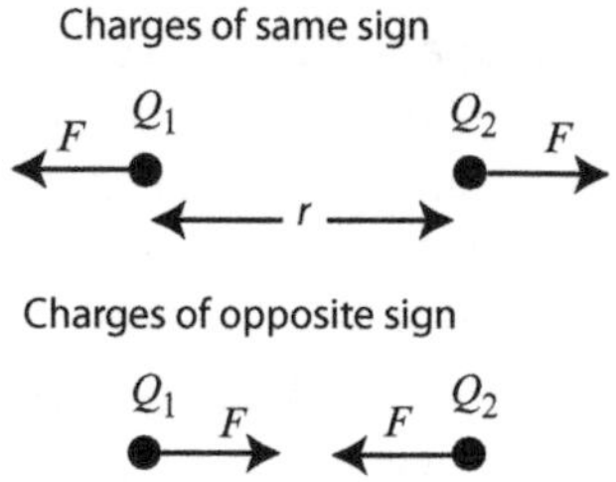

The dependency on $1/r^2$ makes this a so-called 'inverse square law'.

ε_0 is a constant, called **the permittivity of free space**. It also appears in the capacitor equation, $C = \dfrac{\varepsilon_0 A}{d}$, (see Section 2.4).

$\varepsilon_0 = 8.85 \times 10^{-12}\,\text{C}^2\,\text{m}^{-2}\,\text{N}^{-1} = 8.85 \times 10^{-12}\,\text{F m}^{-1}$ (See Quickfire 2)

Charges of same sign

Charges of opposite sign

Fig. 2.6.2 Like charges repel, unlike attract

>> Pointer

Coulomb's law can be written as

$$F = \frac{1}{4\pi\varepsilon_0}\frac{Q_1 Q_2}{r^2}$$

where $\varepsilon_0 = 8.85 \times 10^{-12}$ F m^{-1}.

Note that

$$\frac{1}{4\pi\varepsilon_0} = 8.99 \times 10^{9}\,\text{F}^{-1}\,\text{m}$$

quickfire

② Starting by putting F into more basic SI units, show that

$$\text{F m}^{-1} = \text{C}^2\,\text{m}^{-2}\,\text{N}^{-1}.$$

Example

Two equal compact charges, 0.30 m apart in air, repel each other with forces of 7.8 μN. Calculate the charges.

Answer

If the charge of each is Q, then $\quad F = \dfrac{1}{4\pi\varepsilon_0}\dfrac{Q^2}{r^2}$. Re-arranging:

$$Q = \sqrt{4\pi\varepsilon_0 r^2 F} = \sqrt{4\pi \times 8.85 \times 10^{-12}\,\text{C}^2\,\text{m}^{-2}\,\text{N}^{-1} \times (0.30\,\text{m})^2 \times 7.8 \times 10^{-6}\,\text{N}}$$

$Q = 8.8$ nC Both charges are positive or both are negative.

E due to a point charge

In Coulomb's law we can choose to put $Q_1 = Q$ and to regard this charge as the source of an electric field. Q_2 can be regarded as a test charge, q.

So $F = \dfrac{1}{4\pi\varepsilon_0} \dfrac{Qq}{r^2}$, that is $\dfrac{F}{q} = \dfrac{1}{4\pi\varepsilon_0} \dfrac{Q}{r^2}$

So E at distance r from Q is $E = \dfrac{1}{4\pi\varepsilon_0} \dfrac{Q}{r^2}$.

The direction of the field near Q is radially outwards (as shown) if Q is positive, radially inwards if Q is negative.

Example

An E–r graph (full line) is plotted for point source, Q. Find Q.

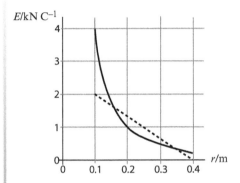

Fig. 2.6.3 E–r for point charge

Answer

Using the graph point at $r = 0.10$ m...

$Q = 4\pi\varepsilon_0 r^2 E$

$\quad = 4\pi \times 8.85 \times 10^{-12}\ \text{C}^2\ \text{m}^{-2}\ \text{N}^{-1} \times (0.10\ \text{m})^2 \times 4000\ \text{N C}^{-1}$

$\quad = 4.45\ \text{nC}$

>> **Pointer**

Note that the presence of the electric field due to Q doesn't depend on q being in place to test for it!

quickfire

③ Note the inverse square law as displayed in Fig. 2.6.3. What happens to E each time r is doubled?

quickfire

④ Calculate E at 0.30 m from a charge of 4.45 nC.

≫ Pointer

To see why the area under the E–x graph gives the pd, regard it as the sum of areas of very thin vertical strips. Over each one of these, E hardly changes, so the strip area really does approach the value $E\Delta x$ as the strips approach zero width.

2.6.3 Electric potential difference and electric potential

These concepts apply only to electric fields arising from stationary charges. We're dealing with *electrostatics*.

(a) Electric potential difference

Consider just a small distance, Δx, moved by a test charge q in the direction of E (Fig. 2.6.4). Then...

Change in PE of q = −work done on q by field

So $\qquad \Delta(PE) = -\text{Force on } q \times \Delta x$

So $\qquad \dfrac{\Delta(PE)}{q} = -\dfrac{(\text{Force on } q)}{q} \times \Delta x$

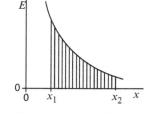

Fig. 2.6.4 work done by E on q

In other words $\quad \Delta V = -E\Delta x$

ΔV, defined as $\dfrac{\text{change in electrical PE of } q}{q}$, is called the **potential difference (pd)** between A and B.

We say that A is at a *higher potential* than B.

UNIT: J C⁻¹ = volt (V)

Suppose E varies with x, for example as in Fig. 2.6.5. To find the potential difference between points x_1 and x_2, some distance apart, we can't usually apply $\Delta V = -E\Delta x$ in one go, because E changes as we go between x_1 and x_2. In this case:

ΔV = area under E–x graph between x_1 and x_2.

Fig. 2.6.5 Area under E–x graph

Example

Using Fig. 2.6.3, estimate the difference in potential between points at 0.10 m and 0.040 m from a charge of 4.45 nC.

Answer

By eye, the (triangular) area under the broken line is *roughly* the same as that under the curve.

So $\quad \Delta V = \frac{1}{2}\text{base} \times \text{height} = \frac{1}{2} \times 0.3\,\text{m} \times 2000\,\text{V m}^{-1} = 300\,\text{V}.$

The point at 0.10 m will be at the higher potential (as the field will do work on a positive test charge going from 0.10 m to 0.40 m).

(b) Electric potential

The **electric PE** of q at point P in an electric field is defined by...

PE of q = work done by field on q as q goes from P to infinity.

The electric **potential** (V or V_E) at P is defined as $V = \dfrac{(PE\ of\ q\ at\ P)}{q}$, so

$$V = \dfrac{\left(\begin{array}{c}\text{work done by the field on a charge } q \\ \text{as } q \text{ goes from P to infinity}\end{array}\right)}{q}$$ Scalar. UNIT: J C^{-1} = volt (V)

- Infinity' means very far from the charge(s) causing the field.
- According to the definition of V, if P is *at* infinity the potential is zero. Defining V as the pd between infinity and P (which is what we've done!) has set up the *convention* that V is zero at infinity.
- Just as E is *force* per unit test charge, so V is the PE per unit test charge and electrical PE of $q = qV$

(c) Electric potential due to a point charge, Q

The *potential* at distance r from a point charge Q is equal to the area under the E–r graph from r to infinity. Using mathematics (integration), this is found to be:

$$V = \dfrac{Q}{4\pi\varepsilon_0 r}$$

According to this equation, as r becomes very large, V approaches zero, in agreement with the convention mentioned above.

V is plotted against r in Fig. 2.6.6. The potential energy of q at distance r from Q is qV, so

$$E_p = \dfrac{Qq}{4\pi\varepsilon_0 r}$$

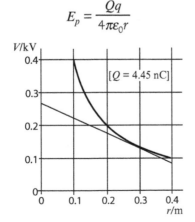

Fig. 2.6.6 V–r for point charge

Although it's often useful to think of Q as 'source charge' and q as 'test charge', we could just as well see two separated charges as a 'system' with PE. The PE equation in the margin suits either point of view.

》Pointer

Put values of the charges along with their signs into the equations

$$V = \dfrac{Q}{4\pi\varepsilon_0 r},\ E_p = \dfrac{Qq}{4\pi\varepsilon_0 r}$$

which will then give V and/or E_p with correct sign(s).

》Pointer

$V = \dfrac{Q}{4\pi\varepsilon_0 r}$ implies that V is *inversely proportional* to r. Compare curve shape in Fig. 2.6.6 with that in Fig. 2.6.3 (E against r).

quickpire

⑥ Calculate the *highest* speed reached by the ion in the example.

quickpire

⑦ Find E at 0.20 m from a charge of 4.45 nC by using a tangent to the $V–r$ graph Fig. 2.6.6. (You shouldn't need to *draw* the tangent!)

Check your answer by simply reading E at $r = 0.20$ m from Fig. 2.6.3.

Example

An ion of charge $+3.20 \times 10^{-19}$ C and mass 4.66×10^{-26} kg is released from rest at a point A, 0.10 m from a charge, Q, of $+10$ nC. It moves further and further from Q. Calculate:

(a) the ion's electrical potential energy (PE) at A

(b) the ion's electrical PE at a point B, 0.30 m from Q

(c) the ion's speed at B.

Answer

(a) Potential at A $= V_A = \dfrac{1}{4\pi\varepsilon_0}\dfrac{Q}{r_A} = 8.99 \times 10^9 \times \dfrac{10 \times 10^{-9}}{0.10\,\text{m}}$ V $= 899$ V

PE of ion at A $= q \times V_A = 3.2 \times 10^{-19}$ C $\times 899$ V $= 2.88 \times 10^{-16}$ J

(b) Potential due to Q is inversely proportional to distance from Q, so PE of ion at B $= \frac{1}{3} \times 2.88 \times 10^{-16}$ J $= 0.96 \times 10^{-16}$ J

(c) $(\text{KE} + \text{PE})$ of ion at A $= (\text{KE} + \text{PE})$ of ion at B

So $\quad 0 + 2.88 \times 10^{-16}$ J $= \frac{1}{2}mv^2 + 0.96 \times 10^{-16}$ J

So $\quad v = \sqrt{\dfrac{2 \times (2.88 - 0.96) \times 10^{-16}\,\text{J}}{m}} = \sqrt{\dfrac{2 \times 1.92 \times 10^{-16}\,\text{J}}{4.66 \times 10^{-26}\,\text{kg}}}$

$\quad = 9.1 \times 10^4$ m s^{-1}

(d) Electric potential gradient

We can write the relationship $\Delta V = -E\Delta x \quad$ as $\quad E = -\dfrac{\Delta V}{\Delta x}$

$\dfrac{\Delta V}{\Delta x}$ is called the potential gradient. Its units are V m^{-1} ($=$ N C^{-1}).

The equation implies that minus the gradient of the tangent to an $E–x$ graph (or, for a radial field, an $E–r$ graph) at a point gives E at that point.

Example

Find E at a distance of 0.30 m from a charge of 4.45 nC, using Fig. 2.6.6.

Answer

Having drawn the tangent shown,

$E = -\dfrac{\Delta V}{\Delta x} = -\dfrac{(0.27 - 0.09)\,\text{kV}}{(0 - 0.40)\text{m}} = -(-0.45\,\text{kV m}^{-1}) = 0.45\,\text{kV m}^{-1}$

E is positive, so the field is in r direction: radially outwards.

2.6.4 E and V due to a number of charges

(a) Electric field strength, E

We find E at a point, **P**, due to a number of nearby point charges, by vector addition of the field strengths at **P** due to individual charges.

Example

Determine the electric field strength at **P** in Fig. 2.6.7(a).

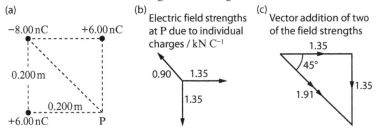

Fig. 2.6.7 Electric field strength (and potential) at P due to three charges

Answer

First calculate the magnitudes of E at **P** due to the individual charges.

For each 6 nC, $E = \dfrac{1}{4\pi\varepsilon_0}\dfrac{Q}{r^2} = 8.99 \times 10^9\,\text{N}\,\text{C}^{-2}\text{m}^2\,\dfrac{6.00 \times 10^{-9}\,\text{C}}{(0.200\,\text{m})^2}$

$= 1.35\,\text{kN}\,\text{C}^{-1}$.

For the -8 nC, $E = 8.99 \times 10^9\,\text{N}\,\text{C}^{-2}\text{m}^2\,\dfrac{8.00 \times 10^{-9}\,\text{C}}{(0.200\,\text{m})^2 + (0.200\,\text{m})^2} = 0.90\,\text{kN}\,\text{C}^{-1}$

The *directions* of these fields at point P are shown in Fig. 2.6.7 (b).

We now add together the fields due to the 6 nC charges. This is done in the vector diagram (Fig. 2.6.7 (c)), noting that $\sqrt{1.35^2 + 1.35^2} = 1.91$.

So the total E at P, now including that from the -8 nC charge, is

$1.91\,\text{kN}\,\text{C}^{-1}$ South East + $0.90\,\text{kN}\,\text{C}^{-1}$ North West = $1.0\,\text{kN}\,\text{C}^{-1}$ South East.

(b) Electric potential, V

This is much easier to calculate than E, because potential is a scalar quantity, so we just take potentials at a point **P** due to individual charges and add them as *numbers*, taking account of any minus signs!

Example

Determine the potential at point **P** in Fig. 2.6.7 (a)

quickfire

⑧ Charges of +0.80 nC are placed at two of the vertices (corners), A and B, of an equilateral triangle of side length 0.070m. Calculate the electric field strength at the third vertex, C. Give its direction as well as magnitude.

quickfire

⑨ Repeat the last Quickfire, but with the charge at A being +0.80 nC, and the charge at B being −0.80 nC.

quickfire

⑩ Calculate the potential at C in Quickfire 8.

quickfire

⑪ Calculate the potential at C in Quickfire 9.

Answer

First calculate the potentials at **P** due to each of the individual charges.

For each 6 nC, $V = \dfrac{1}{4\pi\varepsilon_0} \dfrac{Q}{r} = 8.99 \times 10^9 \text{ N C}^{-2} \text{ m}^2 \dfrac{6.00 \times 10^{-9} \text{ C}}{0.200 \text{ m}} = +270 \text{ V}.$

For the −8 nC, $V = 8.99 \times 10^9 \text{ N C}^{-2} \text{m}^2 \dfrac{-8.00 \times 10^{-9} \text{ C}}{\sqrt{(0.200 \text{ m})^2 + (0.200 \text{ m})^2}} = -254 \text{ V}.$

So total potential at **P** = 2×270 V + (−254 V) = 286 V

2.6.5 Electric field lines and equipotentials

(a) Electric field lines

Key Term

An **electric field line** is a line whose direction at each point along it gives the direction of the field at that point.

These – defined in **Key terms** – are the full lines with arrowheads on them, in the diagrams below. (The broken lines are explained later.)

For 'isolated' charges, the field lines are radial:

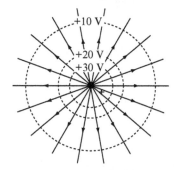
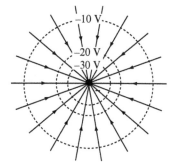

Fig. 2.6.8 Field lines and equipotentials for 'isolated' charges

» Pointer

You won't be expected to draw Fig. 2.6.9 from memory.

- The lines start on positive charges and end on negative charges (even though the diagrams above show only one end of each line).

- The lines have the 'bonus' property of indicating the magnitude of the field *strength*: the closer they come together the stronger the field.

- Electric field lines never intersect. Suppose, for example, we move two 'isolated' charges together. At each point there won't be two separate fields, but a single resultant field – found by vector addition of the fields due to the individual charges.

Grade boost

To boost confidence, choose an off-centre point in Fig 2.6.9. Consider the magnitudes and directions of the field strengths due to the individual charges, and sketch a vector addition diagram. Does the direction of the resultant field agree roughly with that of the field line?

» Pointer

It should be clear that the potential of the middle equipotential in Fig. 2.6.9 is zero.

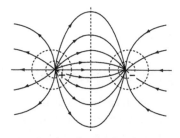

Fig. 2.6.9 Equal and opposite charges

(b) Equipotentials

These are defined in **Key term**. In Figs 2.6.8 & 2.6.9 they are shown (in section) as broken lines. These are rather like contour lines on a map, joining points of equal height.

- Equipotentials and electric field lines cross at right angles.

- In Figs 2.6.8 and 2.6.9, the equipotentials are drawn with roughly equal differences of potential between them. This means that the closer the equipotentials the stronger the field – because the greater the potential gradient (Section 2.6.3 (d)).

> **Key Term**
>
> An **equipotential** is a surface on which all points are at the same potential.

2.6.6 Gravitational field strength, g

A gravitational field is a region where a mass (a so-called *test mass*), m, experiences a force proportional to m.

The **gravitational field strength**, g, at a point is defined as

$$g = \frac{\text{force on } m}{m} \quad \text{that is} \quad g = \frac{F}{m}.$$

- g is a vector quantity. Its units are $\mathrm{N\,kg^{-1}}$ (or – check this! – $\mathrm{m\,s^{-2}}$).
- If a test mass is released in a gravitational field, and other forces on it are negligible, then, using $F = ma$, its acceleration is...

$$a = \frac{F}{m} = \frac{mg}{m} = g.$$

- So, whatever the mass m of the test mass, its free-fall acceleration is equal to the gravitational field strength!

> **Key Term**
>
> The **gravitational field strength**, g, at a point (P) is the force per unit mass experienced by a small 'test' mass placed at P.
>
> $$g = \frac{F}{m} \quad \text{(vector)}$$
>
> UNITS: $\mathrm{N\,kg^{-1}} = \mathrm{m\,s^{-2}}$

Example

When a test mass of $4.00\,\mathrm{kg}$ is hung from a force-meter on the Moon's surface, the reading is $6.50\,\mathrm{N}$. Calculate the gravitational field strength, and state the acceleration of the test mass if allowed to fall.

Answer

$$g = \frac{F}{m} = \frac{6.50\,\mathrm{N}}{4.00\,\mathrm{kg}} = 1.63\ \mathrm{N\ kg^{-1}} \quad \text{Acceleration} = 1.63\ \mathrm{m\ s^{-2}}$$

>> *Pointer*

Newton's law of gravitation doesn't apply accurately in regions of very high *g*, (such as inside, or close to, stars).

quicKfire

⑫ Two electrons are 1.00 mm apart. Calculate:
   a) the electric repulsive force between them
   b) the gravitational force between them.
   ($e = 1.60 \times 10^{-19}$ C, $m_e = 9.11 \times 10^{-31}$ kg)

>> *Pointer*

The last Quickfire should convince you of why gravity is said to be a very weak force.

quicKfire

⑬ The mean radius of the Earth is 6.37×10^6 m and $g = 9.81$ N kg^{-1}. Calculate the mass of the Earth.

quicKfire

⑭ Hence calculate the mean density of the Earth.

2.6.7 Newton's law of gravitation

Every particle attracts every other particle with a force proportional to the product of their masses, M_1 and M_2, and inversely proportional to the square of their separation, r.

Mathematically, $F = (-)G\dfrac{M_1 M_2}{r^2}$.

G is called **Newton's gravitational constant** (or 'big G'). By experiment, $G = 6.67 \times 10^{-11}$ N kg^{-2}m^2 = 6.67×10^{-11} kg^{-1}m^{3}s^{-2}

- M_1 and M_2 are always positive. The force is always attractive; this is sometimes shown with a minus sign (see above).
- The law applies to particles ('point masses'). Stars and planets aren't particles, but they *are* nearly *spherically symmetric*: spherical, with mass distributed evenly all round. When Newton added together (as vectors!) the forces that each

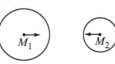

Fig. 2.6.10 spheres attracting

particle of one of these bodies exerts on each particle of another, he found a very neat result: *each body behaves as if all its mass were concentrated at its centre.*

g due to a point, or spherically symmetric, mass

In Newton's law we can choose to put $M_1 = M$, and to regard *M* as the source of a gravitational field. M_2 can be regarded as a test mass, *m*.

So $F = (-)G\dfrac{Mm}{r^2}$ that is $\dfrac{F}{m} = (-)G\dfrac{M}{r^2}$ so $g = (-)\dfrac{GM}{r^2}$.

This is the field strength outside a spherically symmetric object of mass *M*, at distance *r* from its centre. *g* is directed *towards* the centre of *M*; as sometimes shown with the minus sign (above).

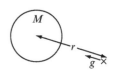

Fig. 2.6.11 *g* outside sphere

Example

The moon's radius is 1.74×10^6 m and $g = 1.62$ N kg^{-1} on its surface. Calculate its mass.

Answer

$$M = \frac{r^2 g}{G} = \frac{(1.74 \times 10^6 \,\mathrm{m})^2 \times 1.62 \,\mathrm{N\,kg^{-1}}}{6.67 \times 10^{-11} \,\mathrm{N\,kg^{-2}\,m^2}} = 7.4 \times 10^{22} \,\mathrm{kg}$$

2.6.8 Gravitational PE, potential and potential difference

The definitions and equations in this section are very similar to those for electric fields in Section 2.6.3, though presented more briefly. Beware the (compulsory) minus signs in certain equations.

(a) Gravitational PE and gravitational potential

The **PE** of test mass m at point P in a gravitational field is defined by...

PE of m = work done by field on m as m goes from P to infinity.

It is always negative because the force (from bodies causing the field) is in the opposite direction to the journey of the mass

The **gravitational potential**, V or V_g, at a point P in a g-field is defined by...

$$V = \frac{\left(\begin{array}{c}\text{work done by the field on a mass } m \\ \text{as } m \text{ goes from P to infinity}\end{array}\right)}{m}$$ Scalar.   UNIT: $J\,kg^{-1} = m^2\,s^{-2}$

- 'Infinity' means very far from the bodies that cause the field.
- The definition keeps to the convention that V is zero at infinity.
- Except at infinity, V will always be negative.
- V is the gravitational potential energy per unit test mass, just as g is the *force* per unit test mass, and gravitational PE of $m = mV$.

(b) Gravitational potential outside a spherically symmetric body

This is given by $V = -\dfrac{GM}{r}$ [minus sign essential]

The PE of m at a distance r from M is $E_p = mV = -\dfrac{GMm}{r}$.

Compare the $g-r$ and $V-r$ graphs (Fig. 2.6.12) for a star of mass $6.00 \times 10^{30}\,kg$. Note that we're including the minus sign in the equation for g, as g is in the opposite direction to that in which r increases.

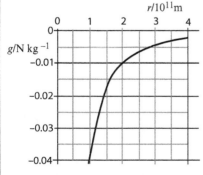 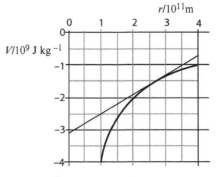

Fig. 2.6.12 $g-r$ and $V-r$ graphs for a star of mass $6.00 \times 10^{30}\,kg$

quickpire

⑮ Using data from the graphs in Fig. 2.6.12, calculate:
   a) the field strength
   b) the potential,
at $8.0 \times 10^{11}\,m$ from a star of mass $6.00 \times 10^{30}\,kg$.

≫ **Pointer**

Note the minus signs in the equations

$$V_g = -\frac{GM}{r}$$
and
$$E_p = -\frac{GMm}{r}$$

≫ **Pointer**

See the *electric* fields section for more detailed explanations.

>> **Pointer**

If m is taken a small distance h against a g field, $\Delta x = -h$, so
$\Delta(PE) = mgh$
This familiar equation applies, for example, to height rises above the Earth's surface, but only if
$h \ll r_{earth}$. It won't usually be the right equation to use for questions on gravitation!

>> **Pointer**

In the example, g at $r = 2.5 \times 10^{11}$ m is simply equal to $\dfrac{V}{r}$ at that radius.
This works only if the field is due to a single body with spherical symmetry. The tangent method is general.

(c) Gravitational potential difference and potential gradient

When a test mass, m, moves a small distance Δx in the same direction as the local gravitational field, it loses PE equal to the amount of work done on it, so

$$\Delta(PE) = -mg\Delta x$$

Dividing both sides of the equation by m, we see that the **gravitational potential difference**, ΔV or ΔV_g, defined by $\dfrac{\Delta(\text{PE of } m)}{m}$, over a small distance Δx in the direction of g, is:

$$\Delta V = -g\Delta x \quad \text{so} \quad g = -\frac{\Delta V}{\Delta x}.$$

Even when we move far enough for g to vary significantly, changes in gravitational potential are represented by areas under the g–x (or g–r) graph, and g at a point is the gradient of the V–x (or V–r) graph at that point (the '**gravitational potential gradient**').

Example

Use the V–r graph in Fig. 2.6.12 to find the gravitational field strength at 2.5×10^{11} m from the centre of a star of mass 6.00×10^{30} kg.

Answer

Using the tangent drawn,

$$g = -\frac{\Delta V}{\Delta r} = -\frac{[-0.7 - (-3.1)] \times 10^9 \, \text{J kg}^{-1}}{[4.0 - 0] \times 10^{11} \, \text{m}} = -6.0 \times 10^{-3} \, \text{N kg}^{-1}$$

This agrees with the value read from the g–r graph.

2.6.9 Conservation of energy in a gravitational field

When spacecraft, satellites and planets move in gravitational fields we can apply the principle of conservation of energy in the form:

$$(\text{gravitational PE} + \text{KE})_1 = (\text{gravitational PE} + \text{KE})_2$$

in which the subscripts refer to any points in the journey.

Example

The moon's mass is 7.35×10^{22} and its radius is 1.74×10^6 m. Calculate how far a rocket will rise if launched vertically from the moon's surface at a speed of $2000 \, \text{m s}^{-1}$.

quickfire

⑯ What other form of energy (apart from kinetic and gravitational potential) should (ideally) be taken into account for bodies travelling through the Earth's atmosphere?

Answer

The furthest point reached, at r_2, let's say, from the centre of the moon, will be when the rocket has run out of KE.

But $\quad$ (PE + KE) just after launch = (PE + KE) at furthest point

So $$-m\frac{GM}{r_1} + \tfrac{1}{2}mv_1^2 = -m\frac{GM}{r_2} + 0$$

Dividing through by the rocket mass, m, and putting in figures, omitting units:

$$-\frac{6.67 \times 10^{-11} \times 7.35 \times 10^{22}}{1.74 \times 10^6} + \tfrac{1}{2}2000^2 = -\frac{6.67 \times 10^{-11} \times 7.35 \times 10^{22}}{r_2} + 0$$

So $$-2.82 \times 10^6 + 2.00 \times 10^6 = -\frac{4.90 \times 10^{12}}{r_2}$$

So $$r_2 = \frac{-4.90 \times 10^{12}}{-0.82 \times 10^6} = 6.0 \times 10^6 \text{ m}$$

And now the final step – easily forgotten... The distance the rocket rises from the moon's surface is $r_2 - r_1 = (6.0 - 1.7) \times 10^6\,\text{m} = 4.3 \times 10^6\,\text{m}$.

Escape speed

This is defined in **Key term**.

'Minimum speed' (in the definition) implies that the body doesn't have any KE left at infinity. The PE will also be zero at infinity by convention.

Example

Calculate the escape speed for a body on the Earth, given that the Earth's mass is $5.97 \times 10^{24}\,\text{kg}$ and its mean radius is $6.37 \times 10^6\,\text{m}$.

Answer

We have $\quad$ (PE + KE) just after launch = (PE + KE) at infinity

So $$-m\frac{GM}{r_1} + \tfrac{1}{2}mv_{esc}^2 = +0 + 0$$

So $$v_{esc} = \sqrt{\frac{2GM}{r_1}} = \sqrt{\frac{2 \times 6.67 \times 10^{-11}\,\text{N kg}^{-2}\,\text{m}^2 \times 5.97 \times 10^{24}\,\text{kg}}{6.37 \times 10^6\,\text{m}}}$$

$$= 11.2 \text{ km s}^{-1}$$

quickfire

⑰ Calculate the escape speed for a body on the Moon.

(mass: $7.35 \times 10^{22}\,\text{kg}$, radius: $1.74 \times 10^6\,\text{m}$.)

quickfire

⑱ Calculate the ratio $\dfrac{\text{escape speed from Jupiter}}{\text{escape speed from Mars}}$.

Use these data:

$M_J = 1.90 \times 10^{27}\,\text{kg}$

$M_M = 6.42 \times 10^{23}\,\text{kg}$

$r_J = 69.9 \times 10^6\,\text{m}$

$r_M = 3.39 \times 10^6\,\text{m}$

Key Term

The **escape speed** (loosely, **escape velocity**) is the minimum speed needed for a body to escape as far as we please ('to infinity') from a point on the surface of a star, planet or satellite.

2.6.10 *g* and *V* due to more than one body

The resultant field strength at a point is the *vector* sum of field strengths due to each body. The resultant potential is the *scalar* sum of potentials due to each body.

Example

(a) Comment on the sketch-graphs of field strength and potential along the line joining Earth (E) and Moon (M), and locate point P.

(Mass of Earth = 5.97×10^{24} kg, mass of Moon = 7.35×10^{22} kg, distance EM = 3.82×10^8 m.)

(b) Calculate the potential at P.

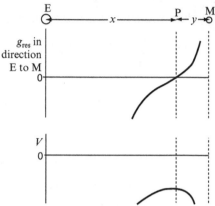

Fig. 2.6.13 *E* and *V* between Earth and Moon

Answer

(a) Between E and P, the Earth's field (towards E) predominates, but, between P and M, the Moon's predominates. P is where the vector sum of the fields is zero,

So $\dfrac{GM_M}{y^2} - \dfrac{GM_E}{x^2} = 0$,   that is $\dfrac{M_M}{y^2} = \dfrac{M_E}{x^2}$.

So $\dfrac{x}{y} = \sqrt{\dfrac{M_E}{M_M}} = \sqrt{\dfrac{5.97 \times 10^{24} \text{ kg}}{7.35 \times 10^{22} \text{ kg}}} = 9.01$ that is $x = 9.01\,y$

But $x + y = 3.82 \times 10^8$ m

so $10.01y = 3.82 \times 10^8$ m and $y = \dfrac{3.82 \times 10^8 \text{ m}}{10.0}$

So $y = 0.38 \times 10^8$ m, and $x = 9.01y = 3.44 \times 10^8$ m

As for the gravitational potential, we know it is always negative and that it will be most negative near E and M. The maximum (least negative) value will be at P because $g = 0$ here, so $\dfrac{\Delta V}{\Delta r} = 0$ at P (the tangent to the V–r graph is horizontal)

(b) The potential at P is the scalar sum of potentials due to E and M.

So $V = -\dfrac{GM_E}{x} + \left(-\dfrac{GM_M}{y}\right)$

$= -6.67 \times 10^{-11} \text{ N m}^2 \text{ kg}^{-2} \times \left(\dfrac{5.97 \times 10^{24} \text{ kg}}{3.44 \times 10^8 \text{ m}} + \dfrac{7.35 \times 10^{22} \text{ kg}}{0.38 \times 10^8 \text{ m}}\right)$

$= -1.29$ MJ

2.6.11 Summary table adapted from the Physics specification

ELECTRIC FIELDS	GRAVITATIONAL FIELDS
Electric field strength, E, at a point is the force per unit charge on a small positive test charge placed at the point.	Gravitational field strength, g, at a point is the force per unit mass on a small positive test mass placed at the point.
Inverse square law for the force between two electric charges $F = \dfrac{1}{4\pi\varepsilon_0}\dfrac{Q_1 Q_2}{r^2}$ (Coulomb's law) $\dfrac{1}{4\pi\varepsilon_0} = 9.0 \times 10^9 \text{ m}^{-1}\text{F}$ is acceptable	Inverse square law for the force between two (point) masses $F = G\dfrac{M_1 M_2}{r^2}$ (Newton's law of gravitation)
F can be attractive or repulsive $E = \dfrac{1}{4\pi\varepsilon_0}\dfrac{Q}{r^2}$ is the field strength due to a point charge in free space (a vacuum) or air	F is attractive only $g = \dfrac{GM}{r^2}$ is the field strength due to a point mass
Potential, V or V_E at a point is the work done per unit charge in bringing a test charge from infinity to that point. $V_E = \dfrac{1}{4\pi\varepsilon_0}\dfrac{Q}{r}$ $PE = \dfrac{1}{4\pi\varepsilon_0}\dfrac{Q_1 Q2}{r}$	Potential, V or V_g at a point is the work done per unit mass in bringing a test mass from infinity to that point. It is always negative. $V_g = -\dfrac{GM}{r}$ $PE = -G\dfrac{M_1 M_2}{r}$
For a charge, q, in an electric field due to stationary charges $\Delta(PE) = q\Delta V_E$	For a mass, m, in a gravitational field $\Delta(PE) = m\Delta V_g$
Field strength at a point is given by $E = -$slope of V_E–r graph at that point	Field strength at a point is given by $g = -$slope of V_g–r graph at that point

1. Three point charges are fixed in the positions shown in Fig. 2.6.14 (a).

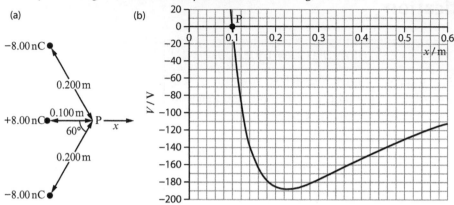

Fig. 2.6.14 Three charges and graph of potential against distance

(a) Calculate from Fig. 2.6.14 (a):

   (i) the field strength at **P**

   (ii) the initial acceleration of a particle of charge $+3.33 \times 10^{-15}$C and mass 3.50×10^{-9} kg released from rest at **P** (ignoring any gravitational effects)

   (iii) the potential at **P**.

(b) Fig. 2.6.14 (b) shows how the potential varies along a portion of the horizontal line through **P** in Fig. 2.6.14 (a). Distance x is measured from the $+8$ nC charge.

   (i) Determine from the graph the value (call it x_0) of x at which E is zero, stating your reason briefly.

   (ii) Explain briefly how the graph shows that E is in the $+x$ direction for $x < x_0$ but in the $-x$ direction for $x > x_0$.

   (iii) What can be said about E when $0.3\,\text{m} \leq x \leq 0.4\,\text{m}$?

   (iv) Explain why, for very large x (well beyond the range of the graph), the potential is nearly given by

   $$V = \frac{Q}{4\pi\varepsilon_0 x} \text{ in which } Q = -8.00\,\text{nC}.$$

(c) Describe how the velocity of the particle in (a)(ii) changes during its journey after release. Forces other than electrostatic are negligible. Calculations aren't required, but your answers to (a)(iii) and (b) are relevant.

2. Newton showed that, according to the inverse square law, a uniform spherical shell (a hollow sphere like a table-tennis ball) would produce a gravitational field that is:
   - at all points inside it: zero
   - at all points outside it: the same as if all its mass were concentrated at its centre.

   (a) Write expressions for the *potential* due to a spherical shell of mass M and radius r_s at a distance r from its centre for
      (i) $r \geq r_s$
      (ii) $r < r_s$ (Hints: you know V when $r = r_s$, and you know that $\Delta V = -g\Delta r$.)

   (b) Think of a uniform solid sphere (of radius a and mass M) as nested spherical shells, like an onion. Show that *inside* the sphere ($r < a$) the field strength due to the sphere at r from its centre is $g_r = \dfrac{r}{a} g_a$
   in which $g_a = (-)\dfrac{GM}{a^2}$.

   (Hint: You might like to assume a density, ρ, which you can get rid of at the end, by expressing it in terms of M and a.)

3. The Sun's mass is 1.99×10^{30} kg and its mean radius is 6.96×10^8 m.
   (a) Show that the escape speed from the Sun's surface is 620 km s^{-1}.
   (b) Calculate the furthest distance that a body could go if it were ejected from the Sun's surface with a speed of 310 km s^{-1}.
   (c) Over 200 years ago the Reverend John Michell suggested that even light, which he thought of as particles or 'corpuscles', would not be able to escape from extremely dense stars. Use the same escape speed equation you used in (a) to find a value for the minimum radius the Sun could have for sunlight still to escape the Sun's gravitational field.

   [Note that the *derivation* of the equation is not valid for light, nor for the extremely large values of field strength that apply, but the equation does happen to give the right value for the Schwarzchild radius, the radius below which a star has to shrink in order to become a black hole.]

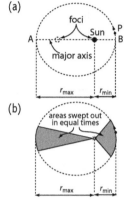

Fig. 2.8.1 Kepler's laws

2.8 Orbits and the wider universe

We summarise how Newton's law of gravitation can be applied to satellites orbiting planets, planets orbiting stars and stars orbiting vast quantities of matter in galaxies. On the way, we recall the idea of 'dark matter' and the use of Doppler techniques to measure the 'radial' velocities of stars. Section 2.8 ends with a brief treatment of Hubble's law and what it seems to tell us about the age of the universe.

2.8.1 Kepler's laws of planetary motion

The laws were discovered by *observation*.

1. Each planet moves in an ellipse, with the Sun at one focus.
2. The line from Sun to planet sweeps out equal areas in equal times.

   This implies that, in its orbit, the closer a planet is to the Sun the faster it moves.

3. For the different planets in the Solar system, the period[2] is proportional to the semi-major axis[3].

   i.e. $\quad T^2 \propto r^3$

   where $r = \dfrac{r_{max} + r_{min}}{2}$ = the semi-major axis.

Newton showed that the planets would move according to just these laws if there were an attractive, inverse-square law, force from the Sun.

2.8.2 Circular orbits in a gravitational field

Most planets' orbits are not very 'eccentric': they're quite close to circular – as is the Moon's orbit, and those of artificial satellites.

We shall now concentrate on circular orbits (which are a special case, for which $r_{min} = r_{max}$, of elliptical orbits).

Fig. 2.8.2 shows a body of mass m orbiting a body of mass M. We assume $M \gg m$, so M remains stationary.

M's gravitational pull provides m with its centripetal acceleration. We apply $F = ma$ to find an equation for T, in two equivalent ways (revise Section 1.5!) ...

Fig 2.8.2 Circular orbit

$$\frac{GMm}{r^2} = mr\omega^2 \qquad \text{or,} \qquad \frac{GMm}{r^2} = m\frac{v^2}{r}$$

Dividing through by m: $\qquad\qquad\qquad$ Dividing through by m:

$$\frac{GM}{r^2} = r\omega^2 \qquad \text{or,} \qquad \frac{GM}{r^2} = \frac{v^2}{r}$$

But $\quad \omega = \dfrac{2\pi}{T}\quad$ so $\quad \dfrac{GM}{r^2} = r\dfrac{4\pi^2}{T^2}\qquad$ Re-arranging: $v^2 = \dfrac{GM}{r}$

But $\qquad T^2 = \dfrac{4\pi^2}{GM}r^3 \qquad$ But $\; v = \dfrac{2\pi r}{T}\quad$ so $\quad T = \dfrac{2\pi r}{v}$

$$\text{So } T^2 = \frac{4\pi^2 r^2}{v^2} = \frac{4\pi^2}{GM}r^3$$

So, at least in the case of circular orbits, we have shown that Kepler's third law follows from the law of gravitation.

(a) Using orbit theory for satellites and planets

Example 1

A geostationary satellite is a man-made satellite with a circular orbit in the plane of the equator, of period 24 hours, so that it is always above one point on the equator. Starting from Newton's law of gravitation, determine the height of the satellite above the Earth's surface. [Mass of Earth = 5.974×10^{24} kg, radius of Earth = 6.37×10^6 m]

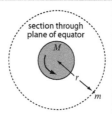

Fig 2.8.3 Satellite in circular orbit

Answer

The Earth's gravitational pull supplies centripetal force on satellite,

so $\qquad \dfrac{GMm}{r^2} = mr\omega^2 \qquad$ that is $\quad GM = r^3\omega^2$

But $\quad \omega = \dfrac{2\pi}{T}\quad$ so $\quad$ that is $\quad GM = r^3\dfrac{4\pi^2}{T^2}\quad$ that is $\quad r^3 = \dfrac{GMT^2}{4\pi^2}$

So $\quad r = \sqrt[3]{\dfrac{GMT^2}{4\pi^2}} = \left(\dfrac{6.67 \times 10^{-11} \times 5.97 \times 10^{24} \times (24 \times 3600)^2}{4\pi^2}\right)^{\frac{1}{3}}$

$\qquad = 42.2 \times 10^6$ m

Finally: height above Earth = $r - 6.34 \times 10^6$ m = 35.8×10^6 m.

Example 2

The distances of the Earth and Jupiter from the Sun are 1.50×10^{11} m and 7.78×10^{11} m. Calculate Jupiter's year in Earth-years.

Answer

We use Kepler's third law. There is no need to evaluate the constant, nor to convert years to seconds – if we do a little algebra first.

$$\frac{T^2}{r^3} = \text{constant} \quad \text{so} \quad \frac{T_J^2}{r_J^3} = \frac{T_E^2}{r_E^3} \quad \text{so} \quad \frac{T_J^2}{T_E^2} = \frac{r_J^3}{r_E^3} = \left(\frac{r_J}{r_E}\right)^3$$

Thus $\quad T_J = T_E\left(\dfrac{r_J}{r_E}\right)^{\frac{3}{2}} = 1.00 \text{ year} \times \left(\dfrac{7.78 \times 10^{11}\,\text{m}}{1.50 \times 10^{11}\,\text{m}}\right)^{\frac{3}{2}} = 11.8 \text{ year}$

> **Pointer**
>
> We can write the T–r relationship as
> $$T = \frac{2\pi}{\sqrt{GM}}r^{3/2}$$

② The moon's orbital radius is 3.84×10^8 m and its orbital period is 27.3 day, Calculate a figure for the mass of the Earth.

quicKfire

③ Use data on the Earth from **Example 2** to find a figure for the mass of the Sun.

> **Pointer**
>
> Square rooting means raising to the power of $\frac{1}{2}$; cube rooting means raising to the power of $\frac{1}{3}$ that is $\sqrt[2]{a} = a^{\frac{1}{2}}$ and $\sqrt[3]{a} = a^{\frac{1}{3}}$ and, for example, $\sqrt{a^3} = a^{\frac{3}{2}}$.

>> Pointer

The equation $v = \sqrt{GM/r}$ for the rotation speed of a particle, m, in the galaxy assumes a spherically symmetric distribution of mass between m and the galactic centre. This is not quite right, because the galaxy, though bulgy near the middle, becomes more disc-like further out.

(b) Rotation of galaxies and dark matter

A spiral galaxy is a huge collection of stars, together with gas and dust, distributed roughly in disc formation and rotating about its centre.

We can use Newton's law of gravitation to predict the rotation speed of a particle of mass m at distance r from the centre of the galaxy.

$$\frac{GMm}{r^2} = m\frac{v^2}{r} \qquad \text{that is} \qquad v^2 = \frac{GM}{r} \qquad \text{and} \qquad v = \sqrt{\frac{GM}{r}}$$

M is the mass of material between the particle and the galactic centre. We've made the crude assumption of spherical symmetry (see Pointer).

Astrophysicists can estimate the observable, so-called 'baryonic' mass, M_B, of the galaxy from the electromagnetic (e-m) radiation it gives out, and from the e-m radiation absorbed by gas clouds. Most of this mass is concentrated in the 'central bulge' of the galaxy.

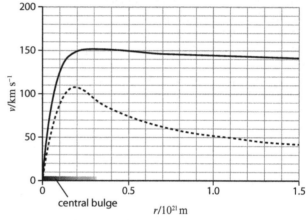

Fig 2.8.4 Galactic rotation curves for NGC3198: observed (full line) and 'expected' (dotted line)

This means that *outside* the central bulge, we'd expect M to change little and to be roughly equal to M_B. Hence the theoretical graph (dotted), which, *outside the central bulge*, is drawn to follow $v = \sqrt{GM/r}$ with a constant value for M equal to M_B as estimated for the galaxy NGC3198.

Example

Determine the value of M on which the dotted graph is based.

Answer

Selecting the point at $r = 0.5 \times 10^{21}$ m:

$$M = \frac{rv^2}{G} = \frac{0.50 \times 10^{21}\,\text{m} \times (74 \times 10^3\,\text{m s}^{-1})^2}{6.67 \times 10^{-11}\,\text{N kg}^{-2}\text{m}^2} = 3.0 \times 10^{45}\,\text{kg}$$

quickfire

④ Why do the $v-r$ graphs rise from zero to their maximum values?

quickfire

⑤ What can be said about rv^2 for the dotted (theoretical) graph beyond about 0.3×10^{21} m?

quickfire

⑥ Evaluate the ratio
$$\left(\frac{v_{\text{solid line}}}{v_{\text{broken line}}}\right)^2$$
at 1.5×10^{21} m and state the significance of this ratio (making the crude assumption of spherical symmetry).

For some galaxies we can *measure v* at various distances, *r*, using a *Doppler shift* technique. (See Section 2.8.4.) The solid line on the graph is the *measured v* plotted against *r* for NGC3198. We conclude that

- The mass in the galaxy extends far beyond the central bulge (as *v* does not fall off with *r* in the expected way).
- The overall mass of the galaxy is greater than M_B by something like a factor of 10.

These discrepancies are far too large to be accounted for by errors of *measurement*. Instead, the most favoured theory is that galaxies contain **dark matter**, which, unlike baryonic matter, doesn't interact with electromagnetic waves, and is therefore 'invisible' to astronomers whatever wavelength of e-m radiation they are examining, whether they are looking for radiation being given out or being absorbed.

The nature of dark matter is still a mystery. For many years WIMPs were the favoured constituents. These are *weakly interacting massive particles* predicted by theories of particle physics involving 'supersymmetry', but their existence is proving suspiciously hard to verify by experiment. A more recent suggestion is that dark matter may arise from certain decay modes of the famous Higgs boson...

> **Key Term**
>
> **Dark matter** is matter which we can't detect by any sort of e-m radiation, but whose existence we infer from its gravitational effect.

2.8.3 Binary systems

Our system will consist of two bodies, of masses M_1 and M_2 (for example two stars or a star and a large planet) in which each mass is large enough to affect the other's motion detectably.

Both M_1 and M_2 will orbit the same point, C, called the system's **centre of mass**, with the same angular velocity. We shall assume *circular* orbits. The position of C is given (see Fig 2.8.5) by

$$r_1 = \frac{M_2}{M_1 + M_2}d \quad \text{and} \quad r_2 = \frac{M_1}{M_1 + M_2}d$$

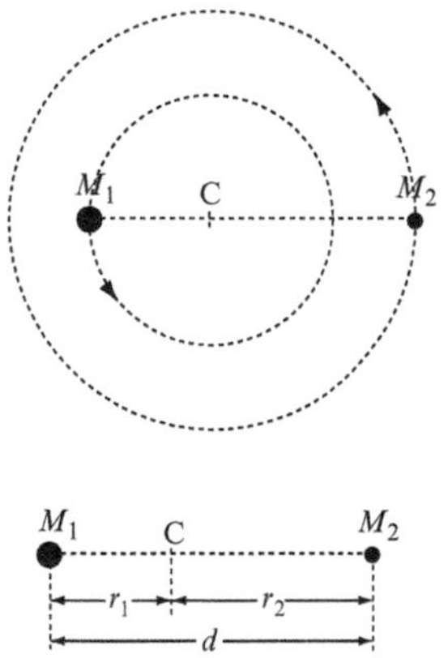

Quick checks to make on these equations

- $r_1 + r_2 = d$.
- $M_1 r_1 = M_2 r_2$ since $M_1 r_1 \omega^2 = M_2 r_2 \omega^2$

(This simply equates the magnitudes of mass × acceleration for M_1 and M_2 – since they exert equal and opposite forces on each other.)

Fig 2.8.5 Binary system

> **» Pointer**
>
> If $M_2 \ll M_1$, C may be inside M_1, as is the case for the Earth-Moon system.

Kepler's third law for binary systems.

Both bodies have a periodic time of $T = 2\pi \sqrt{\dfrac{d^3}{G(M_1 + M_2)}}$.

> **» Pointer**
>
> The speeds of the bodies are in the same ratio as their orbital radii, since
> $$\frac{v_1}{v_2} = \frac{r_1 \omega}{r_2 \omega} = \frac{r_1}{r_2}$$

≫ Pointer

Here's how we obtain

$$T = 2\pi\sqrt{\frac{d^3}{G(M_1 + M_2)}}.$$

Apply $F = ma$ to M_1 and divide both sides of the equation by M_1, giving

$$\frac{GM_2}{d^2} = r_1\omega^2.$$

Then express r_1 in terms of d and ω in terms of T and re-arrange. Go on: do it!

quicKfire

⑦ Two stars, each with the same mass as the Sun, form a binary system (circular orbits) of period 1.00 Earth year. Calculate d_{SS}/d_{SE} in which d_{SS} is the separation of the centres of the stars, and d_{SE} is the same as the Earth's orbital radius. [Use the Kepler 3 equation for the new system and for the Earth.]

Example

A double star (Delta Capricorni A and B) has an orbital period of 8.83×10^4 s (just over a day). The orbital speeds of A and B are $9.25 \times 10^4 \, \mathrm{m\,s^{-1}}$ and $20.6 \times 10^4 \, \mathrm{m\,s^{-1}}$. Calculate the masses of A and B.

Answer

First we'll work out r_A, and r_B, using the speeds and the period.

$$v_A = \frac{2\pi r_A}{T}, \text{ so } r_A = \frac{v_A T}{2\pi} = \frac{9.25 \times 10^4 \ \mathrm{m\,s^{-1}} \times 8.83 \times 10^4 \mathrm{s}}{2\pi} = 1.30 \times 10^9 \mathrm{m}$$

and

$$r_B = \frac{20.6 \times 10^4 \ \mathrm{m\ s^{-1}} \times 8.83 \times 10^4 \ \mathrm{s}}{2\pi} = 2.89 \times 10^9 \ \mathrm{m}$$

Now we'll find $M_A + M_B$ by re-arranging the 'Kepler 3' equation.

$$M_A + M_B = \frac{4\pi^2 d^3}{T^2 G} = \frac{4\pi^2 (1.30 \times 10^9 \mathrm{m} + 2.89 \times 10^9 \mathrm{m})^3}{(8.83 \times 10^4 \mathrm{s})^2 \times 6.67 \times 10^{-11} \mathrm{N\,kg^{-2}\,m^2}} = 5.58 \times 10^{30} \mathrm{kg}$$

But

$$r_A = \frac{M_B}{M_A + M_B} d \quad \text{so} \quad M_B = (M_A + M_B)\frac{r_A}{d}$$

$$= 5.58 \times 10^{30} \mathrm{kg} \frac{1.30 \times 10^9 \ \mathrm{m}}{4.19 \times 10^9 \ \mathrm{m}}$$

Thus $M_B = 1.7 \times 10^{30} \mathrm{kg}$ and $M_A = (5.58 - 1.74) \times 10^{30} \mathrm{kg} = 3.9 \times 10^{30} \mathrm{kg}$

2.8.4 Measurements of velocity by Doppler shift

This has been the key technique in recent discoveries of planets outside the solar system (exoplanets), black holes and extra mass in galaxies.

(a) The Doppler effect for light

If a source of light and an observer are moving further apart, the observed wave frequency is lowered (as each successive wave 'peak' has further to travel). Since $\lambda = c/f$, the observed wavelength is increased; we say there is a 'Doppler shift to the red'. If source and observer are moving closer there is a 'shift to the blue'.

Fig 2.8.6 Star and observer moving further apart

The change, $\Delta\lambda$, in wavelength of a spectral line is given by $\dfrac{\Delta\lambda}{\lambda} = \dfrac{v_{rad}}{c}$, in which λ is the unshifted wavelength and v_{rad} is the **radial velocity**: the component along our line-of-sight of the star's (or other source's) velocity relative to us. A source approaching us is given a negative value of v_{rad}, so $\Delta\lambda$ correctly comes out negative. (The equation is approximate, but when $v_{rad} \ll c$, as in the cases below, it is almost exact.)

(b) Doppler determination of a star's orbital velocity

Our example will be the star 51 Pegasi (51 Peg). Lines in its spectrum were found to have a regularly changing Doppler shift. For one absorption line, identified as a hydrogen line of laboratory wavelength 656.281 nm, the extreme shifts are -7.36486×10^{-11} m and -7.38982×10^{-11} m. So the extreme radial velocities are

$$v_{max} = c\frac{\Delta\lambda}{\lambda} = 2.99792 \times 10^{8}\,\mathrm{m\,s^{-1}} \times \frac{-7.36486 \times 10^{-11}\,\mathrm{m}}{656.281\,\mathrm{m}} = -33\,643\,\mathrm{m\,s^{-1}}$$

$$v_{min} = c\frac{\Delta\lambda}{\lambda} = 2.99792 \times 10^{8}\,\mathrm{m\,s^{-1}} \times \frac{-7.36982 \times 10^{-11}\,\mathrm{m}}{656.281\,\mathrm{m}} = -33\,757\,\mathrm{m\,s^{-1}}$$

In fact, the radial velocity, plotted against time varies as in Fig 2.8.7(a).

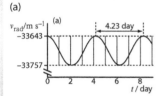
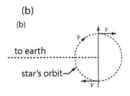

Fig 2.8.7 Radial velocity of 51 Peg plotted against time

A sinusoidal variation in v_{rad} is just what we'd find for a star in circular orbit. For orbital speed v, assuming that the orbit is being viewed edge-on (Fig 2.8.7(b)), $v_{max} = v$ and $v_{min} = -v$, so $v_{max} - v_{min} = v - (-v) = 2v$. The sinusoid on the graph is displaced 'downwards', showing that, in addition to the orbital motion, there is a large steady velocity component towards us. But it will still be true that $v_{max} - v_{min} = 2v$. So, for 51 Peg:

Orbital speed $= \frac{1}{2}(v_{max} - v_{min}) = \frac{1}{2}(33\,757 - 33\,643)\,\mathrm{m\,s^{-1}} = 57\,\mathrm{m\,s^{-1}}$

2.8.5 Determining a binary system from one body's orbit

If, from Doppler measurements, we discover a star in orbital motion, and are able to measure its orbital velocity, v, and period T, then as long as we can estimate the star's mass (probably from the radiation it emits), we can

Grade boost

Don't forget to give the values of v_{min} and v_{max} their right signs in the orbital velocity equation
$$v = \tfrac{1}{2}(v_{min} - v_{max})$$

quickfire

⑧ Calculate the mean radial velocity of 51 Peg, which is the radial velocity of the centre of mass of the 51 Peg system.

quickfire

⑨ Light of wavelength 510 nm from the star Tau Boötes is found to exhibit a varying Doppler shift between -291.5×10^{-13} m and -275.5×10^{-13} m, with a period of 3.31 days. Calculate the star's orbital speed and orbital radius.

calculate the mass and orbital radius of its companion! This is best shown by an example (with commentary).

Example

For the star 51 Pegasi, $v = 57.0\,\text{m s}^{-1}$, $T = 4.23$ day (see Section 2.8.4). From its spectrum and luminosity 51 Pegasi is estimated to be a Sun-like star of mass $M_S = 2.1 \times 10^{30}$ kg. Calculate the mass and orbital radius of its companion.

Answer

1. First we determine the star's orbital radius using $2\pi r_s = vT$

$$r_s = \frac{vT}{2\pi} = \frac{57.0\,\text{m s}^{-1} \times 4.23 \times 86\,400\,\text{s}}{2\pi} = 3.32 \times 10^6\,\text{m}$$

This is about half the Earth's radius. A small stellar orbit like this, often called a 'wobble', betrays the presence of a companion, making up a binary system. The companion of 55 Pegasi is invisible and (judging by its small effect on the star) of very much smaller mass than the star – presumably a planet.

2. We can now find the separation, d, of star (S) and planet (P), using

$$T = 2\pi \sqrt{\frac{d^3}{G(M_S + M_P)}} \quad \text{so} \quad d = \left[\frac{T^2 G(M_S + M_P)}{4\pi^2}\right]^{\frac{1}{3}}.$$

As we've argued, the planet is in this case of much smaller mass than the star ($M_P < 0.01\,M_S$), so putting M_S instead of $M_S + M_P$ in the equation will make hardly any difference.

Thus $d = \left[\dfrac{T^2 G M_S}{4\pi^2}\right]^{\frac{1}{3}}$

$$= \left[\frac{(4.23 \times 86\,400)^2 \times 6.67 \times 10^{-11} \times 2.1 \times 10^{30}\,\text{m}^3}{4\pi^2}\right]^{\frac{1}{3}}$$

So $d = 7.8 \times 10^9\,\text{m}$. But $r_S = 3.32 \times 10^6\,\text{m}$, so $r_P = d - r_S = 7.8 \times 10^9\,\text{m}$.

So, to 2 sf, r_P is indistinguishable from d (and would be even with 3 sf).

3. We can now find the planet's mass, M_P, using $M_P r_P = M_S r_S$:

$$M_P = M_S \frac{r_S}{r_P} = 2.1 \times 10^{30}\,\text{kg} \times \frac{3.32 \times 10^6\,\text{m}}{7.8 \times 10^9\,\text{m}} = 8.9 \times 10^{26}\,\text{kg}.$$

This is about half the mass of Jupiter. (The real figure will be larger if we are not, in fact, viewing the system nearly edge on.) The orbital radius, r_P, is about a seventh that of Mercury – so the planet will be roasting! This was the first discovery (1995) of a planet belonging to a star similar to the Sun.

2.8.6 Deep space and Hubble's law

As we've seen (Section 2.8.2 (b)) stars are orbiting within galaxies. Galaxies themselves are clustered and orbiting within the clusters! But superimposed on these 'local' movements there seems to be a motion *away from us*. The greater the distance, D, of the object away from us, the greater its red shift, so according to $\dfrac{\Delta\lambda}{\lambda} = \dfrac{v}{c}$, the greater its velocity, v, away from us, that is its *radial velocity* of recession.

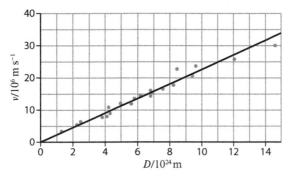

Fig 2.8.8 Recession velocity, v, against distance, D

As Fig 2.8.8 shows, if we plot v against D we get a straight line through the origin. This gives us Hubble's law (see **Terms and definitions**).

Where does the data come from for the graph? D can be determined for objects of known luminosity (so-called *standard candles*) by measuring the power per m^2 reaching us from them (see **Quickfire 12** for an example). v is determined by Doppler shift measurements.

The age of the universe

Just because objects are moving away *from us* doesn't mean that we have a special central place. Suppose that everything started moving away from a central origin at the time of the Big Bang. Then everything will be moving away from everything else (see second **Pointer**).

Using the symbols introduced above, at time t after the Big Bang, $D = vt$. But Hubble's law, $v = H_0 D$ can be written as $D = \dfrac{1}{H_0} v$, so, – cue fanfare – we deduce that,

Age of universe, $t = \dfrac{1}{H_0} = \dfrac{1}{2.20 \times 10^{-18}\ \text{s}^{-1}}$

$= 4.55 \times 10^{17}\ \text{s}$

$= 14.4 \times 10^9\ \text{year}$

Key Term

Hubble's law
Each object has a superimposed radial velocity, v, given by

$$v = H_0 D$$

in which D is its distance from us. H_0 is a constant called the Hubble constant.

$H_0 = 2.20 \times 10^{-18}\ \text{s}^{-1}$

>> *Pointer*

Astronomers give H_0 as (approximately) $68\ \text{km s}^{-1}\ \text{Mpc}^{-1}$, in which the parsec (pc) is a unit of distance: $1\ \text{pc} = 3.09 \times 10^{16}$ m. Show that this gives $H_0 = 2.20 \times 10^{-18}\ \text{s}^{-1}$

quickfire

⑫ Type 1A supernovas are standard candles with peak luminosity of 1.6×10^{36} W. Calculate the distance away of SN1937c, whose observed peak intensity was $8.2 \times 10^{-12}\ \text{W m}^{-2}$.

>> *Pointer*

Here we use bold print for vectors. Suppose we're moving at velocity $\mathbf{v}_0$ from a central point. Our displacement in time t is $\mathbf{r}_0 = \mathbf{v}_0 t$. For some other object, X, $\mathbf{r}_X = \mathbf{v}_X t$.

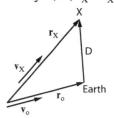

So $\mathbf{r}_X - \mathbf{r}_0 = (\mathbf{v}_X - \mathbf{v}_0)t$. That is displacement of X *from us* = velocity of X *relative to us* $\times t$.

This value, the reciprocal of H_0, is called the *Hubble time*. Equating it to the age of the universe assumes that the expansion rate has always been the same. In fact we believe that the rate has slowed owing to gravitational forces acting 'inwards'. The expansion rate, and the value of the Hubble 'constant' would have been higher in the past (indeed H_0 is sometimes called the present value of the *Hubble parameter, H*). Our current best estimate for the age of the universe is 13.8×10^9 year, which accords with H once being higher than H_0, but is based on detailed models of the early evolution of the universe.

quickfire

⑬ Calculate ρ_c from the equation, $\rho_c = \dfrac{3H_0{}^2}{8\pi G}$ being sure to show that the units work out correctly.

quickfire

⑭ How many hydrogen atoms would be needed per m^3 to give a mean density of ρ_c?

>> *Pointer*
For comparison with your value for ρ_c, the density of residual air in an 'ultra-high vacuum' in industrial equipment is reckoned to be about $10^{-9}\,kg\,m^{-3}$.

2.8.7 Critical density of the universe

Imagine a sphere of radius r drawn around our galaxy, a sphere large enough to contain millions of other galaxies, so, treating it as a homogeneous sphere with mean density, ρ, its mass, M, will be:

$$M = \text{volume} \times \text{density} = \tfrac{4}{3}\pi r^3 \rho$$

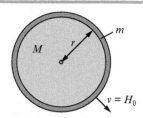

Fig. 2.8.9 Expanding shell

Now consider a thin 'shell' (mass m) of universe surrounding the sphere. Will this shell, moving away from us at speed $v = H_0 r$, have enough kinetic energy to keep moving outwards, against the gravitational pull of the sphere?

As we saw in Section 2.6.9, for the escape of a body of mass m to infinity with no spare KE,

$$\text{Initial KE of body} = \text{PE at infinity} - \text{Initial PE (at } r\text{)}$$

So $\qquad \tfrac{1}{2}mv^2 = 0 - \left(-\dfrac{GMm}{r}\right) \qquad$ that is $\qquad v^2 r = 2GM$

Substituting expressions for v and M given above,

$$(H_0 r)^2 r = 2G\tfrac{4}{3}\pi r^3 \rho$$

Cancelling through by r^3 and re-arranging,

$$\rho_c = \dfrac{3H_0{}^2}{8\pi G}$$

We've stuck the subscript 'c' on to ρ, as the value of ρ given by this formula is called *the critical density* of the universe.

The fate of the universe depends on how its actual mean density, ρ_{actual}, compares with ρ_c. We believe that ...

If $\rho_{\text{actual}} > \rho_c$: expansion slows to zero, then contraction at growing rate.

If $\rho_{\text{actual}} = \rho_c$: expansion will slow to zero rate but only at infinite size.

If $\rho_{\text{actual}} < \rho_c$: expansion will continue for ever.

Evidence suggests that (if we take account of dark matter), the universe is very close to its critical density, that is $\rho_{\text{actual}} \approx \rho_c$. Yet observed deviations from Hubble's law show that we appear to be in an era of accelerating expansion. There is much that we don't fully understand.

The expansion of space

We've presented this section and the last using the Newtonian concept of space as an unchanging background against which objects move and interact. That's fine at A-level, but be aware that we now believe space itself to be expanding, so objects get further apart as space expands. Take the case of red shift... Light from very distant objects was emitted long ago. By the time it reaches us, space has expanded, and with it the light waves – their wavelength has increased. Rather different from the explanation presented earlier, but, we stress, **not** tested at A-level.

1. Dwellers on Mars plan to install a communications satellite which orbits Mars so that it is always above a fixed point on the planet's surface.

    (a) Use one of Kepler's laws to explain why the orbit must be circular rather than elliptical.

    (b) Determine the height of the satellite above the planet's surface.

    (c) State what other restriction there is on the position of the satellite's orbit.

    Martian data: Mass = 6.42×10^{23} kg; radius = 3.37×10^6 m; day = 24.7 (earth) hours

2. A satellite of mass m is in circular orbit of radius r about a planet of mass M.

    (a) Express the orbital speed of the satellite in terms of M and r and hence show that the satellite's kinetic energy is given by

    $$E_k = \frac{GMm}{2r}$$

    (b) Because of collisions with particles, the satellite gradually loses energy and spirals inwards, that is the radius r of its orbit decreases. Using (a), state what happens to the satellite's kinetic energy as r decreases.

    (c) Explain why your answer to (b) does not violate the principle of conservation of energy. [Consider the total energy of the satellite.]

3. (a) (i) Show that the speed, v, of an object in a circular orbit of radius r around a spherically symmetric body of mass M is given by

    $$v = \sqrt{\frac{GM}{r}}$$

    (ii) Hence derive an equation relating $\dfrac{v_1}{v_2}$ for two objects orbiting the same body to the ratio of their orbital radii, r_1 and r_2.

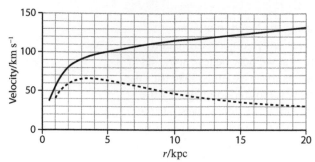

(b) The solid line in the graph shows how the measured rotation speed of matter in the galaxy M33 varies with its distance, r, from the centre of the galaxy. r is expressed in kiloparsec (kpc); $1\,\text{kpc} = 3.086 \times 10^{19}\,\text{m}$.

The broken line is the curve that would be expected if objects were orbiting known 'baryonic' matter (which mainly lies within about 4 kpc of the centre of the galaxy).

(i) Check whether the broken line is consistent with the relationship you derived in (a) (ii). Take $r_1 = 5\,\text{kpc}$ and $r_2 = 15\,\text{kpc}$.

(ii) Explain why we would not expect the relationship to hold for $r_1 = 2\,\text{kpc}$ and $r_2 = 6\,\text{kpc}$.

(iii) Choosing a point beyond $r = 5\,\text{kpc}$ estimate the central mass M on which the broken line is based. Express M in terms of $M_\odot$, the Sun's mass; $M_\odot = 1.99 \times 10^{30}\,\text{kg}$.

(c) By comparing the solid line with the broken line, deduce what you can about the actual mass of material in the galaxy, and its distribution. Further calculations are not wanted.

4. A cleaned-up radial velocity curve is given for the star HD11964, showing the 'wobble' due to one of its planets. The star's mass is estimated as $2.24 \times 10^{30}\,\text{kg}$ (that is $1.13\,M_\odot$).

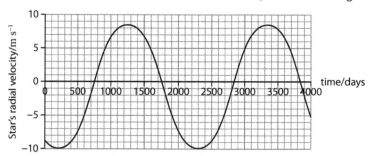

(a) Determine from the curve:

(i) the period of the orbit and hence the separation of the planet from the star, stating the approximation you are making

(ii) the star's orbital velocity, hence the radius of its orbit (assumed circular). Deduce the planet's orbital radius.

(b) Calculate the planet's mass.

(c) Calculate the planet's orbital speed.

(d) Judging by its mass and orbital radius, which planet of the solar system does this planet of HD11964 most resemble? Consult a table of solar system data.

Component 2 Summary
Electricity and the universe

2.4 Capacitance

- Structure of a parallel plate capacitor
- The capacitor as a store of energy and separated charge
- Definition of capacitance, $C = \dfrac{Q}{V}$
- Factors affecting capacitance; $C = \dfrac{\varepsilon_0 A}{d}$ and the qualitative effect of the dielectric
- Electric field in a capacitor, $E = \dfrac{V}{d}$
- Energy stored, $U = \frac{1}{2}QV = \frac{1}{2}CV^2 = \frac{1}{2}\dfrac{Q^2}{C}$
- Capacitors in series and parallel
  $$\frac{1}{C} = \frac{1}{C_1} + \frac{1}{C_2} + \ldots; \; C = C_1 + C_2 + \ldots$$
- Charging and discharging a capacitor
  $$Q = Q_0\left(1 - e^{-\frac{t}{RC}}\right); \; Q = Q_0 e^{-\frac{t}{RC}}$$
  where RC is the time constant

2.6 Electric and gravitational fields

- The features of gravitational and electric fields given in the table in Section 2.6.11
- The gravitational field outside a spherically symmetric body is the same as if all the mass were concentrated at the centre
- The gravitational (or electric) field at a point is the force per unit mass (charge) on a small test mass (charge) placed at that point and is a vector
- Field lines give the direction of the field at that point; the field due to spherically symmetric masses (charges)
- The definition of gravitational (electric) potential; the scalar nature of potential
- How to calculate the resultant field and potential at a point due to a number of point masses (charges)
- The applicability and use of the equation $\Delta U_\text{P} = mg\Delta h$

2.8 Orbits and the wider universe

- Kepler's laws of planetary motion
- Newton's law of gravitation,
  $F = G\dfrac{M_1 M_2}{r^2}$ and its application to orbits
  including derivation of Kepler's 3rd law,
  $T^2 = \dfrac{4\pi^2}{GM}r$ for circular orbits
- Using orbital data to calculate the mass of the central object
- The rotation curves of galaxies and the implied existence of dark matter
- The centre of mass of two spherically symmetric objects; mutual orbits
  $$T = 2\pi\sqrt{\frac{d^3}{G(M_1 + M_2)}}$$
- Determining the properties of a binary system from data on one body's orbit
- The Doppler shift, $\dfrac{\Delta\lambda}{\lambda} = \dfrac{v}{c}$, of radiation from objects in mutual orbit (viewed edge on)
- The Hubble relationship, $v = H_0 D$; $\dfrac{1}{H_0}$ as an approximation to the age of the universe
- The derivation of $\rho_\text{c} = \dfrac{3H_0^2}{8\pi G}$ for the critical density of a 'flat' universe

Specified practical work

- Investigation of the charging and discharging of a capacitor to determine the time constant
- Investigation of the energy stored in a capacitor

Component 3 Knowledge and Understanding

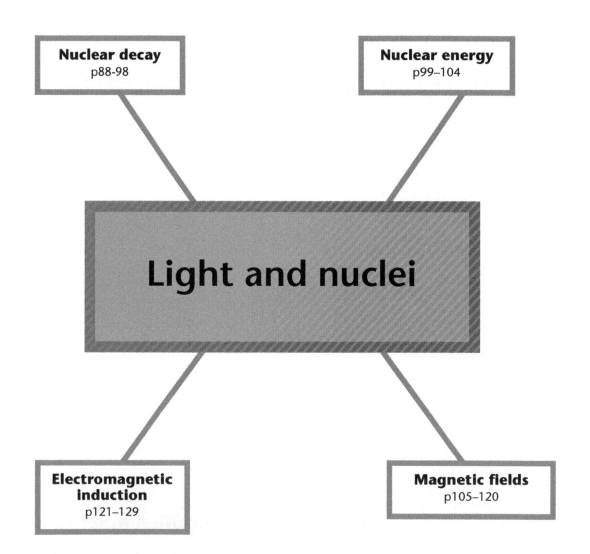

Nuclear decay
p88-98

Nuclear energy
p99–104

Light and nuclei

Electromagnetic induction
p121–129

Magnetic fields
p105–120

Nuclear decay

Nuclear decay as a spontaneous random event; the nature and penetrating power of α, β and γ radiation; handling background radiation; activity and the decay constant; exponential decay and the use of logarithms in nuclear decay calculations.

→ **p88-98** →

Nuclear energy

The meaning and use of $E = mc^2$ and the unified mass unit; calculations of nuclear binding energy; conservation of mass/energy to particle interactions; the importance of binding energy per nucleon in fission and fusion.

→ **p99–104** →

The magnetic field

Forces on moving charges and current-carrying conductors in a magnetic field; the Hall voltage; the fields due to straight wires and solenoids; forces between current-carrying conductors; the motion of charged particles in uniform electric and magnetic fields; particle accelerators.

→ **p105–120** →

Electromagnetic induction

The concepts of magnetic flux and flux linkage; the laws of Faraday and Lenz in electromagnetic induction; emf generated by a moving linear conductor and a rotating coil in a magnetic field.

→ **p121–129** →

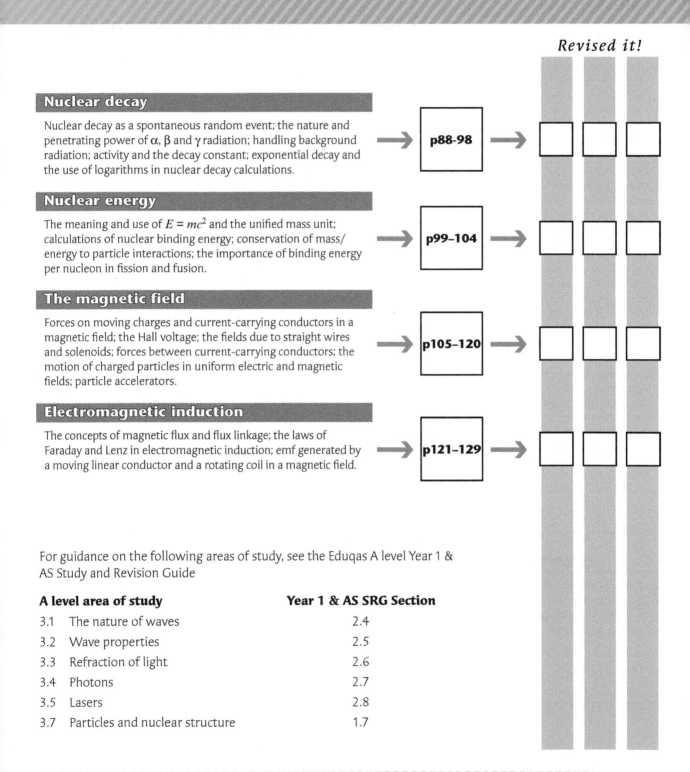

For guidance on the following areas of study, see the Eduqas A level Year 1 & AS Study and Revision Guide

A level area of study	Year 1 & AS SRG Section
3.1 The nature of waves	2.4
3.2 Wave properties	2.5
3.3 Refraction of light	2.6
3.4 Photons	2.7
3.5 Lasers	2.8
3.7 Particles and nuclear structure	1.7

>> **Pointer**

You need to learn the penetration of the three types of radiation.

>> **Pointer**

You also need to know the ionising powers of the three types of radiation but this is just the reverse order of the penetration, i.e. α – highest ionising, β – intermediate ionising and γ – low ionising.

>> **Pointer**

Once you know the penetration and ionising properties you can explain the relative dangers of the radiation i.e. α – most dangerous but only if inside the body (it can't penetrate skin), β – intermediate danger, γ – lowest danger but most difficult to shield against.

>> **Pointer**

In these nuclear reaction equations remember that the A number and Z number are conserved, i.e. the individual totals of the A and Z numbers on the right-hand side (RHS) are equal to the total on the LHS.

quickfire

① Balance the equations by putting in the missing nucleon and proton numbers

$$^{241}_{95}\text{Am} \rightarrow \underline{\quad}\text{Np} + \underline{\quad}\text{He}$$

$$^{7}_{4}\text{Be} \rightarrow \underline{\quad}\text{Li} + \underline{\quad}$$

$$^{99}_{43}\text{Tc*} \rightarrow \underline{\quad}\text{Tc} + \underline{\quad}$$

3.6 Nuclear decay

3.6.1 Alpha (α), beta (β) and gamma (γ)

These three types of radiation are all ionising radiation. They are ionising because they knock out electrons from atoms or molecules. The ionised particles that are produced will be highly reactive and will react with other molecules nearby. In living tissue this can lead to all sorts of damage at the cellular level including damage to DNA, possibly leading to cancer. However, our bodies are subject to attacks from background radiation every minute of the day and life expectancy, remarkably, is no shorter in places with very high background radiation. On the other hand, an absorbed radiation dosage of only 8 J per kilogram is lethal to all humans.

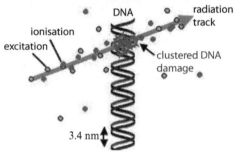

Fig. 3.6.1 The effect of ionising radiation on cell molecules

An alpha (α) particle is a fast-moving helium nucleus (i.e. $^{4}_{2}\text{He}^{2+}$ or $^{4}_{2}\alpha$ but the 2+ is usually omitted). It is more highly ionising than both β and γ radiation. In contrast, the fact that α radiation is so ionising means that it loses its energy very quickly and has low penetration. In fact, the range of α particles is only a few cm in air and they are absorbed by a sheet of paper.

Example of α decay: $^{238}_{92}\text{U} \rightarrow ^{234}_{90}\text{Th} + ^{4}_{2}\text{He}$

A beta (β^-) particle is a fast-moving electron and is usually written as $^{0}_{-1}\text{e}$ or $^{0}_{-1}\beta$. It is more highly ionising than γ-radiation but less so than α radiation. Likewise, β^- has an intermediate penetrating power – it is usually stopped by a few mm of aluminium or a few metres of air.

Example of β^- decay: $^{14}_{6}\text{C} \rightarrow ^{14}_{7}\text{N} + ^{0}_{-1}\beta + ^{0}_{0}\overline{\nu}_e$

Note: Not all β radiation consists of electrons. β^+ radiation consists of positrons (anti-electrons), which are best written as $^{0}_{1}\beta$ (or $^{0}_{1}\text{e}$)when completing nuclear equations.

Example of a β^+decay: $^{39}_{20}\text{Ca} \rightarrow ^{39}_{19}\text{K} + ^{0}_{1}\beta + ^{0}_{0}\nu_e$

γ-radiation is a high-energy, low-wavelength electromagnetic wave or photon that originates from an excited nucleus. It is less ionising than both α and β particles but is consequently more penetrating. γ-radiation is stopped by around 15 cm of lead (Pb) or around a metre of concrete.

Example of γ decay: $^{60}_{28}\text{Ni*} \rightarrow ^{60}_{28}\text{Ni} + ^{0}_{0}\gamma$

The asterisk denotes that the original Ni (nickel) nucleus is in an excited state.

3.6.2 Properties of nuclear radiation

(a) Distinguishing by penetrating power

The relative absorptions of the three nuclear radiation types are represented in Fig. 3.6.2.

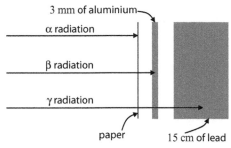

Fig. 3.6.2 Penetrating power of nuclear radiation

This leads nicely to a simple experiment to investigate which type(s) of radiation are present in a radioactive source.

Consider the following set up:

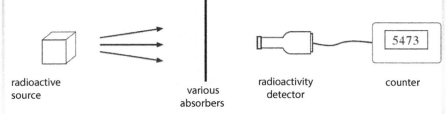

Fig. 3.6.3 Investigating the absorption of radiation

By placing various absorbers between the source and detector you can work out which radiation is emitted by the source. Here are the relevant steps:

Place a sheet of paper between the source and detector. If there's a significant drop in count rate (say from 5473 down to 4000), there must be α radiation present. Then place a piece of aluminium a couple of mm thick between the source and detector. If there's a ***further*** significant drop (say down to 2000), there must be β radiation present also. Whatever signal that's left (above background radiation of ~0.5 counts s^{-1}) must be due to γ radiation. Note that you don't need a γ-absorber to do this experiment because anything significant that's left over after the β absorber must be due to γ radiation. With the above quoted results you would conclude that the count rate due to α radiation is around 1500 Bq, that due to β radiation is around 2000 Bq and that due to γ radiation is around 2000 Bq.

> **》Pointer**
> α, β and γ all have typical energies of the order ~1 MeV and the ionisation energies of atoms and molecules of the order ~10 eV. Hence, each radioactive particle has the energy to produce around 10^5 ions. An α particle produces these ions in a short distance and hence has low penetration but high ionisation, a γ-ray does the opposite – it produces its ~100 000 ions over a large distance and so is said to have low ionisation (even though it produces roughly the same number of ions).

Example

The results in Table 3.6.1 were obtained. Explain which types of radiation are present in the radioactive source.

Absorber	Count rate / s^{-1}
None	8894
3 sheets of paper	5473
None	8921
0.5 mm of Aluminium foil	5455
None	8860
10 cm of lead	56
None	8888

Table 3.6.1 Absorber results (1)

Answer

The count rate drops by about 3500 s^{-1} when 3 sheets of paper are used as an absorber. This is a sure sign that α radiation is present. There is a very similar drop when 0.5 mm of aluminium is used, which suggests that there is no β radiation (i.e. the aluminium doesn't absorb any more than the paper). However, there must be γ radiation present because there's a large count rate after using 0.5 mm of aluminium. Confirmation of the presence of γ radiation is given when the count rate is significant after using 10 cm of lead (it's still detecting 56 s^{-1} which is considerably greater than background radiation).

Some things to beware in this type of data:

1 Paper will absorb some of the β radiation.

2 Aluminium will absorb some of the γ radiation.

3 There is a substantial random error in all radioactive count readings (see the variation in count rate with no absorber).

4 Normally, you'll have to take account of the background radiation (it's around 0.5 counts s^{-1} but that's insignificant in the above results).

When you take all these things into account, you need to be looking for significant drops in count rates and not drops of a few percent as different absorbers are put in place.

(b) Distinguishing by deflection

Moving charged particles are deflected by magnetic fields. If your teacher has a cloud chamber you might well have seen a demonstration of this effect (if not there are plenty of good videos of working cloud chambers on YouTube). Fig. 3.6.4 is a picture of what you might see if you had $^{226}_{88}\text{Ra}$ as the radioactive source and a cloud chamber in a magnetic field ($^{226}_{88}\text{Ra}$ emits all three types of

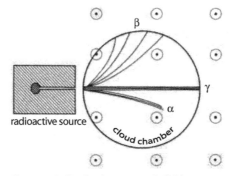

Fig. 3.6.4 Deflection in a magnetic field

nuclear radiation and a cloud chamber is a clever device that gives a vapour trail where ionising radiation has been).

For practice, you should use FLHR to check that the directions of the curvature of the α and β tracks are correct. Also, note that γ radiation doesn't curve because it isn't charged.

Very similar results can be obtained if you use an electric field instead of a magnetic field (Fig. 3.6.5). The shapes of the paths will be slightly different – the magnetic field will give arcs of circles whereas the electric field will give parabolas.

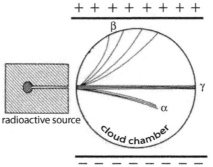

Fig. 3.6.5 Deflection in an electric field

3.6.3 Background radiation

Here's a typical pie chart to explain where we get most of our dosage of ionising radiation.

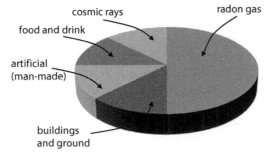

Fig. 3.6.6 Sources of background radiation

》 Pointer

No knowledge of how a cloud chamber works or details of cloud chamber experiments are required but you might be expected to apply what you know about the motion of charged particles to a given set up (a bit like the knowledge required of particle accelerators Section 3.9).

》 Pointer

Remember four things about background radiation:
1 Natural radioactive elements everywhere.
2 Cosmic rays (from Sun and space).
3 Man-made (mainly X-rays).
4 Only around half a count a second.

》 Pointer

You might have to subtract background radiation from count rate readings to obtain true count rates due to the source.

Pointer
You might even have to add background radiation to a source count in order to obtain the actual count produced by the counter.

quicKfire

⑥ A sample of material is tested and 542 counts were recorded in 20 minutes. Explain whether the sample is significantly radioactive. (Background radiation is approximately 0.5 counts s^{-1}.)

Three of the five sources are essentially the same because they come from naturally occurring elements on the Earth. Radon gas that we breathe, food and drink and buildings and the ground are all natural sources coming originally from radioactive elements such as potassium-40, carbon-14, uranium and thorium. Cosmic rays are completely different and mainly arise from high energy particles arriving at the Earth's atmosphere from the Sun. The vast majority of our man-made radiation dosage is a direct result of having an X-ray image taken but a tiny percentage is a result of nuclear power or nuclear weapon testing. Although the sources of nuclear radiation are everywhere and seemingly vast, in nearly all the world a radiation detector will sit quietly randomly clicking at a mean rate of about half a count every second – telling us that there's nothing to fear.

3.6.4 Theory of radioactivity

All the mathematical theory that follows is based on one simple principle – that radioactivity is an entirely random process and depends purely on the number of radioactive nuclei present. Therefore, for a radioactive sample, the number of nuclei disintegrating per second is proportional to the number of nuclei present, i.e.

$$\text{disintegrations per second} \propto N \text{ (the number of nuclei)}$$

We now define a **decay constant** λ (see **Key term**) as the constant of proportionality and λ will depend on the nucleus that's decaying. We also define the activity A as the number of disintegrations per second (see Key term).

$$\text{disintegrations per second} = A = \lambda N$$

But every time a nucleus disintegrates, the number of nuclei decreases, so the number of disintegrations per second is $-\dfrac{\Delta N}{\Delta t}$. The minus sign is there because the number N is decreasing and $\dfrac{\Delta N}{\Delta t}$ must be negative. Hence, the equation that is the foundation of all nuclear decay theory is:

$$\frac{\Delta N}{\Delta t} = -\lambda N$$

If you're doing A-level Maths you should be able to integrate the above equation but from a physics point of view, it's better to understand what the equation means and why the final answer makes sense.

The equation $\dfrac{\Delta N}{\Delta t} = -\lambda N$ tells you that nuclei are being lost at a rate that is proportional to the number of nuclei $\left(\dfrac{\Delta N}{\Delta t} \propto -N\right)$. So, when there are the most nuclei (at the start), you get the greatest activity and the number of nuclei decreases at the highest rate. As the number of nuclei decreases, the activity drops and nuclei are lost at a proportionately lower rate.

If you plot a graph of number of nuclei against time, the number must go from a certain value (N_0 say) to zero. But you also know that the gradient of the line $\left(\dfrac{\Delta N}{\Delta t}\right)$ is always decreasing (because $\dfrac{\Delta N}{\Delta t} \propto -N$). You can probably guess the shape of the graph – it's an exponential decay. This is always true when you have a variable that is decreasing at a rate proportional to the variable itself. In this case, the number of nuclei is decreasing at a rate proportional to the number of radioactive nuclei present. A similar case (to a good approximation) is the flow of water out of a burette – the rate of decrease of height of water is proportional to the height of the water. Similarly, in capacitors, the rate of decrease of charge stored is proportional to the amount of charge held. All these examples give exponential decays.

The equation that gives the time variation of the number of radioactive nuclei is:

$$N = N_0 e^{-\lambda t}$$

This tells you that the number of radioactive nuclei, N, decreases exponentially from an initial number of nuclei. Now, let's multiply the equation by the decay constant i.e.

$$\lambda N = \lambda N_0 e^{-\lambda t}$$

but the activity $A = \lambda N$ and the initial activity $A_0 = \lambda N_0$, hence:

$$A = A_0 e^{-\lambda t}$$

Therefore, the activity, A, also decreases exponentially from an initial value of A_0. These three equations ($N = N_0 e^{-\lambda t}$, $A = \lambda N$ and $A = A_0 e^{-\lambda t}$) appear in the Eduqas Data booklet and you don't need to remember them. However, the syllabus states that you should be able to derive the equation involving the half-life (see **Key term**).

We'll use the definition of the half-life and the equation $A = A_0 e^{-\lambda t}$ to derive an expression for the half-life ($T_{1/2}$). When the time reaches $t = T_{1/2}$ the activity drops from $A_0 = \frac{1}{2}A_0$ (this is the definition of $T_{1/2}$). Putting these values into the equation you get

$$\tfrac{1}{2}A_0 = A_0 e^{-\lambda t_{1/2}} \quad \rightarrow \quad \tfrac{1}{2} = e^{-\lambda t_{1/2}} \rightarrow 2 = e^{\lambda t_{1/2}}$$

and, taking logs:

$\ln 2 = \lambda T_{1/2}$ or $\lambda = \dfrac{\ln 2}{T_{1/2}}$ as the equation appears on the Data booklet.

Example

A sample of carbon-14 has a mass of $150\,g$. Calculate:

(i)   the number of nuclei present (the mass of a carbon-14 atom is $14.00\,u$)

(ii) its decay constant (the half-life of carbon-14 is 5730 year)

(iii) the initial activity of the $150\,g$ sample of carbon-14

(iv) the activity after 2500 years

(v)   the mass of carbon-14 after 11 460 years

(vi) the time for the activity of carbon-14 to decrease to 10% of its initial value.

quickⲪire

⑦ When being used to investigate a radioactive source, a Geiger counter registers 1000 counts in half an hour. Background radiation is 0.35 counts per second. Calculate the true number of decays per second detected by the Geiger counter due to the radioactive source.

Key Term

Half-life = The time taken for the number of radioactive nuclei N (or the activity A) to reduce to half the initial value. Unit: s but often given as minute/hour/day/year.

》 Pointer

You can also derive $\lambda = \dfrac{\ln 2}{T_{1/2}}$ from the equation $N = N_0 e^{-\lambda t}$. You'll need to put in the time for the number of nuclei to drop from N_0 to $\frac{1}{2}N_0$.

》 Pointer

Calculating the number of nuclei is often the least well done of all these parts – there is no equation, you just have to understand the mole or the unified mass unit.

⑧ Uranium-238 has a half-life of 4.47×10^9 year. An initial sample of U238 has a mass of 25.2 kg. Calculate:

(i) the decay constant in s^{-1}

(ii) the number of nuclei of uranium-238 in the initial sample (the mass of a uranium atom is 238 u)

(iii) the initial activity of the sample

(iv) the activity of the sample after 3 half-lives

(v) the activity of the sample after 5.00×10^9 year

(vi) the time (in years) for the activity of the sample to decrease to 30% of its initial value.

Key Term

Radio-isotope = an isotope that is radioactive (remember that isotopes have the same atomic number Z but different mass number A).

⑨ Explain very briefly how you would modify the procedure for the dice experiment if you were by yourself and had around 400 dice.

Answer

(i) mass of atom $= 14.00 \times 1.66 \times 10^{-27} = 2.324 \times 10^{-26}$ kg

Either number of atoms $= \dfrac{0.150}{2.324 \times 10^{-26}} = 6.45 \times 10^{24}$

or number of moles $n = \dfrac{150}{14.00} = 10.71$

∴ number of atoms $= N = nN_A = 10.71 \times 6.02 \times 10^{23} = 6.45 \times 10^{24}$

and this is also the number of nuclei.

(ii) $\lambda = \dfrac{\ln 2}{5730 \text{ year}} = 1.210 \times 10^{-4} \text{ year}^{-1} = 3.84 \times 10^{-12} \text{ s}^{-1}$

(iii) $A = \lambda N = 3.84 \times 10^{-12} \times 6.45 \times 10^{24} = 2.48 \times 10^{13} \text{ Bq}$

(iv) $A = A_0 e^{-\lambda t} = 2.48 \times 10^{13} e^{-1.210 \times 10^{-4} \times 2500} = 1.83 \times 10^{13} \text{ Bq}$

(v) Easier than it looks: 11 460 is 2 half-lives. So the mass will reduce to one-quarter, i.e. mass = 37.5 g

(vi) Trickier – we need to take logs:

$$A = A_0 e^{-\lambda t} \quad \rightarrow \quad \frac{A}{A_0} = e^{-\lambda t} \quad \rightarrow \quad \ln\left(\frac{A}{A_0}\right) = -\lambda t$$

Now: $\dfrac{A}{A_0} = 10\% = 0.1$

∴ $t = \dfrac{\ln 0.1}{\lambda} = -\dfrac{\ln 0.1}{1.210 \times 10^{-4} \text{ year}^{-1}} = 19\,000$ years

3.6.5 Specified practical work

(a) A Dice analogy for nuclear decay

In order to carry out this experiment to a reasonable accuracy a large number of dice is required – around 1000 would be an ideal number for a class of around 10 pupils with each pupil responsible for 100 dice. Before commencing you must decide which number on the dice signifies a decay – a six is as good a number as any but you might have dice that are coloured on one face only.

First you count all your dice – this is the initial number of 'radioactive nuclei'. Then you throw all the undecayed dice. Separate, count and record all the 'decayed' dice from the last throw. Gather together the remaining undecayed dice and throw them again. Repeat this procedure for around ten throws – there should be around 15% of the dice remaining but this is very variable and highly dependent on chance.

The best way of collating the data is in a spreadsheet with each student's results in a column. It's a simple matter then to group together all the data for better results. A typical table of results is shown (Table 3.6.3).

Throw	Number of dice										
	S1	S2	S3	S4	S5	S6	S7	S8	S9	S10	Total
0	100	100	100	100	100	100	100	100	100	100	1000
1	80	83	86	84	83	83	81	81	85	80	826
2	69	71	74	70	72	70	67	69	72	63	697
3	55	60	65	59	62	61	54	56	57	52	581
4	45	47	57	45	52	54	42	46	45	43	476
5	39	37	49	38	43	45	35	41	38	33	398
6	30	28	37	32	34	37	27	36	29	29	319
7	22	24	29	26	30	33	23	28	21	26	262
8	17	20	23	22	24	29	17	22	16	21	211
9	15	15	20	17	17	24	13	16	11	15	163
10	13	12	17	14	15	20	10	14	8	10	133

Table 3.6.3 Typical results for the dice experiment

The last column can be used to plot what should be an exponential decay.

The data shown have been simulated using random numbers and is representative of a 'lucky' good data set. However, a plot of activity against throw number (Fig. 3.6.7) reveals that the data are far from perfect. The 'activity' for each throw can be found by subtracting subsequent total number of dice, e.g. the initial activity is $1000 - 826 = 174$.

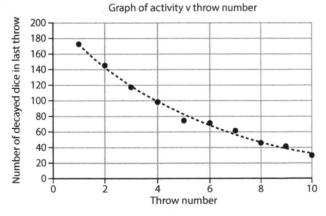

Fig. 3.6.7 Graph of 'activity' against time for the dice experiment

Why is the activity graph so much noisier than the number of dice graph? The simple answer is because the numbers are smaller. When random errors are involved, it really is a case of big is beautiful – the more readings you take, the better your results.

quickfire

⑩ Draw an empty table showing how you would present your data for Quickfire 9.

quickfire

⑪ The uncertainty in a total count of N is $\sqrt{N}$. Explain why the fractional uncertainty decreases as the number of counts increases.

quickfire

⑫ The uncertainty in the value of the activity is the square root of the activity itself. Use this to put error bars on the graph in Fig. 3.6.7.

quickfire

⑬ In light of the error bars that you have added to the graph, evaluate the agreement of the data points to exponential decay theory.

>> **Pointer**

You might expect the decay constant of a dice to be around 0.167 throw^{-1}. Remember that the decay constant is the probability of decay per unit time.
For the dice the probability of decay each throw is $\frac{1}{6}$ or 0.167.

quicKfire

(14) Measure the gradient of the graph in Fig. 3.6.8 and compare its value with 0.167.

>> **Pointer**

Your value in Quickfire 14 should be too large. In order to obtain the real expected value of the gradient you need to solve the following equation (think about it!!)

$$\frac{5}{6} = e^{-\lambda \times 1}$$

giving $\lambda = 0.182$.

>> **Pointer**

The usual gamma source in schools and colleges is Co60. Unfortunately it has a short half-life of about 5 years so it rapidly becomes unusable. A way round this is to use a source which gives out α, β and γ radiation and to shield out the α and β by putting a thin (2–3 mm) aluminium absorber between the source and G-M tube. Ra226 (half-life 1500 years) is commonly used.

There are many ways of analysing these data. The simplest way is to check that the half-life is a constant. In the first graph it takes around 3.5 throws for the number of dice to halve from 1000 to 500. After another 3.5 throws (7 in total), the number of dice has halved again to 250. After another 3 throws (10 in total) the number of dice has nearly halved again which is exactly what would be expected. Hence, the graph seems to be in good agreement with an exponential decay theory.

We can do better by taking logs of the equation

$$N = N_0 e^{-\lambda \times \text{throw number}}$$

giving $\ln N = \ln N_0 - \lambda \times \text{throw number}$

So:

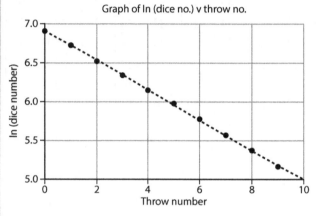

Fig. 3.6.8 Graph of ln (dice no.) against throw number

(b) The variation of intensity of gamma rays with distance

In principle, this is a very straightforward investigation and, as long as the separation of source and detector is large compared with the size of the source, we expect the result to be an inverse-square relationship, i.e.

$$R \propto \frac{1}{d^2}$$

where R is the count rate (corrected for background radiation) and d the separation of the source and detector. The set-up in Fig. 3.6.9 is typical.

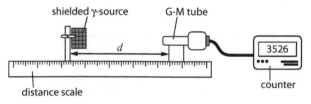

Fig. 3.6.9 Inverse square law for γ rays

The trouble is we cannot measure the actual separation of the source (within the shielding) and the effective position of the detector in the G-M tube) i.e. d is not the true separation. So we have to allow for this.

With the set-up above we therefore expect the count rate R due to the source (i.e. corrected for background) to depend upon d according to:

$$R = \frac{k}{(d - \varepsilon)^2}$$

where k and ε are constants and ε is a correction which allows for the unknown positions.

If we take the square root of the equation and invert it we get:

$$\frac{1}{\sqrt{R}} = \frac{d}{\sqrt{k}} - \frac{\varepsilon}{\sqrt{k}}.$$

Hence, if the inverse square law holds, a graph of $\frac{1}{\sqrt{R}}$ against d will be a straight line. The procedure is quite straightforward and is the subject of Q6 in the Extra questions.

1. Explain which radioisotope from the choices given should be used for the following:
   (a) A radioactive sterilisation suite for ensuring microbe free medical instruments.
   (b) A radioactive source for checking paper thickness.
   (c) A radioactive salt is to be ingested and later enter the kidney where it will kill cells in a small cancer (less than a cm in diameter).
   (d) A radioactive tracer is placed in an underground oil pipe to locate the position of a leak.
   (e) A radioactive source in a smoke detector.
   (f) A radioactive 'poison' is required to 'eliminate' a spy by placing it in a drink.

   Choice of radioisotopes
   A – a radioisotope emitting alpha particles with a short half-life
   B – a radioisotope emitting beta particles with a short half-life
   C – a radioisotope emitting gamma rays with a short half-life
   D – a radioisotope emitting alpha particles with a long half-life
   E – a radioisotope emitting beta particles with a long half-life
   F – a radioisotope emitting gamma rays with a long half-life

2. Explain why beta particles have a much greater curvature in a magnetic (or electric) field than both alpha particles and gamma rays. Hint: you should consider charge, speed and mass.

3. A sample containing cobalt-60 (Co60) has an initial activity of 5.34 GBq. Cobalt-60 has a decay constant of 4.167×10^{-9} s^{-1}. Calculate:

(a) its half -life in years

(b) the mass of Co60 and the number of radioactive atoms in the sample

(c) the activity after 8 years

(d) the activity after 5.22 half-lives

(e) the time taken for the activity to drop to 12.5% of its initial activity (easier than it first appears)

(f) the time taken for the activity to drop to 1.34 GBq.

4 An old tree is unearthed from a peat bog and dated by using carbon dating. The natural ratio of C14 to C12 is 1×10^{-12} but this decreases exponentially after the death of all organisms. C12 does not disintegrate, it is a stable nucleus. The half-life of C14 is 5730 years (this is also the half-life of the ratio of C14 to C12). If the ratio of C14 to C12 in the dead tree is 0.221×10^{-12}, calculate the approximate year (BCE[1]) when it died.

5 Old rocks can be dated by using the ratio of lead to uranium within these rocks. U235 ($^{235}_{92}$U) decays to Pb207 ($^{207}_{82}$Pb) via 7 alpha decays and 4 beta decays.

(a) Show that Pb207 ($^{207}_{82}$Pb) is the expected result from 7 alpha decays and 4 beta decays starting from U235 ($^{235}_{92}$U).

(b) How many alpha decays and how many beta decays will it take for U238 ($^{238}_{92}$U) to decay to Pb206 ($^{206}_{82}$Pb)?

(c) The (overall) half-life of U235 decaying to Pb207 is 700 million years. Assuming that the ratio of Pb207 to U235 starts as zero, explain why the ratio of Pb207 to U235 is 1.00 after one half-life.

(d) How long will it take for the ratio of Pb207 to U235 to reach a value of 3.00?

[Hint: the ratio $\frac{3}{\frac{4}{1}}$ is 3.00.]

(e) Calculate the age of a rock that has a ratio of Pb207 to U235 of 1.35.

(f) The (overall) half-life of U238 decaying to Pb206 is 4.5 billion years. Calculate the ratio of Pb206 to U238 in the rock in part (e).

6. A group of students investigated the inverse square law for gamma radiation. They obtained the following measurements:

Background radiation = 105 counts in 5 minutes

Distance, d / cm	10	15	20	25	30	50	70
Total count	780	301	174	249	255	225	310
Count time / min	1	1	1	2	3	5	10

(a) In terms of uncertainty, suggest why they measured the counts over different times.

(b) Draw up another table of counts per minute, R, corrected for background radiation.

(c) Use a graph of $\frac{1}{\sqrt{R}}$ against d and draw a best fit straight line to

confirm the validity of the inverse square law and calculate the values of k and ε.

[1]BCE = Before Common Era = age in years − 2016 (at time of writing).

3.8 Nuclear energy

3.8.1 Mass-energy equivalence, $E = mc^2$

Perhaps the most famous of all physics equations is:

$$E = mc^2$$

which actually gives a relationship between mass and energy – two seemingly completely different concepts. Nuclear energy is based on this equation and benefits greatly from c^2 being a large number (9×10^{16}). This means that the energy produced when $1\,\text{kg}$ of matter is 'lost' is

$$E = mc^2 = 1 \times (3 \times 10^8)^2 = 9 \times 10^{16}\,\text{J (or } 90\,000\,000\,000\,000\,000\,\text{J})$$

One way of 'losing' $1\,\text{kg}$ of mass is to annihilate $0.5\,\text{kg}$ of antimatter with $0.5\,\text{kg}$ of matter. Unfortunately (or possibly fortunately), $0.5\,\text{kg}$ of isolated antimatter does not exist on Earth and hence, more subtle methods of using nuclear energy must be employed.

The first confirmation of the equation $E = mc^2$ came from Cockcroft and Walton's experiment that also 'split the atom' for the first time in 1932 (they received a Nobel prize in 1951). This is the reaction they used and these are the results that they obtained:

$$^7_3\text{Li} + ^1_1\text{H} \rightarrow ^4_2\text{He} + ^4_2\text{He} + 17.1\,\text{MeV of energy}$$

They bombarded ^7_3Li nuclei with hydrogen nuclei (protons) and obtained two helium nuclei (α particles) along with a considerable sum of extra energy. This extra energy must come from 'lost' mass according to Einstein's equation but did the numbers tie in?

Here are the masses of the nuclei involved presented in a new unit – the unified atomic mass unit ($1\,\text{u} = 1.66 \times 10^{-27}\,\text{kg}$).

mass of $^7_3\text{Li} = 7.0144\,\text{u}$ mass of $^1_1\text{H} = 1.0073\,\text{u}$ mass of $^4_2\text{He} = 4.0015\,\text{u}$

Total mass of LHS $= 7.0144 + 1.0072$ $= 8.0216\,\text{u}$

Total mass of RHS $= 4.0015 + 4.0015$ $= 8.0030\,\text{u}$

i.e. the mass lost $= 0.0186\,\text{u}$

to use $E = mc^2$ you need $1\,\text{u} = 1.66 \times 10^{-27}\,\text{kg}$ from the Data booklet.

$$E = 0.0186 \times 1.66 \times 10^{-27} \times (3 \times 10^8)^2 = 2.779 \times 10^{-12}\,\text{J}$$

To convert from J to eV, you need to divide by e ($1.6 \times 10^{-19}\,\text{C}$)

$$E = \frac{2.78 \times 10^{-12}\,\text{J}}{1.60 \times 10^{-19}\,\text{J eV}^{-1}} = 1.74 \times 10^7\,\text{eV} = 17.4\,\text{MeV}$$

which is very close to Cockcroft and Walton's result of $17.1\,\text{MeV}$.

> **Pointer**
> The word 'lost' is in inverted commas because the mass is not really lost – it is the mass of the energy itself that is released.

> **Pointer**
> 90 PJ is over 10 000 times the energy yield of the bomb dropped on the Japanese city of Hiroshima on 6 August 1945.

> **Key Term**
> The **unified atomic mass unit**, $\text{u} = \frac{1}{12}$ the mass of an isolated $^{12}_6\text{C}$ atom in its ground state.
> $1\,\text{u} = 1.66 \times 10^{-27}\,\text{kg}$ (3 sf)

> **quickpire**
> ① The mass loss in a nuclear reaction is $0.542\,\text{u}$. How much MeV is released in the reaction?
> ($1\,\text{u} \equiv 931\,\text{MeV}$)

> **quickpire**
> ② Convert the following:
> (a) $12.0\,\text{u}$ into kg,
> (b) $401 \times 10^{-27}\,\text{kg}$ into u.

> **Pointer**
> When calculating nuclear energies of reactions always do (LHS mass − RHS mass) × 931 and you have your answer in MeV. Note, for reactions that don't release energy, the answer will be zero or negative.

quickfire

③ How much energy is released in a nuclear reaction when:

(a) $0.666 \times 10^{-9}\,\text{kg}$ of mass is 'lost'?

(b) $0.007\,892\,\text{u}$ of mass is 'lost'?

quickfire

④ Calculate the energy released in the nuclear reaction:

$^6_3\text{Li} + {}^2_1\text{H} \rightarrow {}^4_2\text{He} + {}^4_2\text{He}$

The masses of the nuclei involved are:

mass of ^6_3Li = 6.014 u

mass of ^2_1H = 2.013 u

mass of ^4_2He = 4.002 u

There is, in fact, an easier way of obtaining the correct answer because you can always use the equivalence $1\,\text{u} \equiv 931\,\text{MeV}$. This means that a mass of $1\,\text{u}$ is equivalent to $931\,\text{MeV}$ of energy. So, all you have to do is multiply your mass loss in u by 931 and you get your final answer in MeV i.e. $0.0186 \times 931 = 17.3\,\text{MeV}$ (the slight discrepancy is because all the constants are only given to 3 sf).

The unified mass unit is a particularly useful unit of mass at the atomic or nuclear scale and you will see a lot of it in this section. You will also be converting regularly from u to MeV using $1\,\text{u} \equiv 931\,\text{MeV}$.

3.8.2 What makes some nuclei stable and others unstable?

A complete answer to this question is impossible but at A level you need to be able to explain in terms of binding energy.

First, let's consider the stability of electrons in orbits. There's an attractive force between the nucleus and the electron (+ve and -ve) which holds the electrons in place. The same is true of nucleons in the nucleus. There's an attractive force (strong nuclear force) to hold the nucleons together in the nucleus (it's about 100 times greater than the +ve +ve repulsion of protons).

Whenever an attractive force exists, as the particles come closer they lose potential energy and this is the energy that can be given out. In a chemical reaction, the electrons are more stable in the final products so they've lost potential energy and energy has been given out.

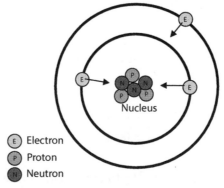

Nucleus

E Electron
P Proton
N Neutron

Fig. 3.8.1 Forces in the atom

It's very similar in nuclear reactions, when the nuclei become more stable they lose PE and give out energy but about a million times more than is given out in chemical reactions.

What happens is that as the particles come closer together their masses and their potential energies decrease. This is true even in chemical reactions – the mass of the system will decrease after an exothermic reaction but this change

Key Term

Binding energy = The energy that has to be supplied in order to separate a nucleus into its constituent nucleons.

Alternatively, it's the energy released (or the decrease in PE) when the constituent nucleons form the nucleus.

UNIT: J or MeV

in mass is difficult to detect. In a nuclear reaction, the change in mass is a million times greater and is easily measured with a mass spectrometer (even in 1932 this was measured quite accurately).

The name given to this change in PE as nucleons are brought together to form a nucleus is **binding energy** and it's defined formally in the **Key term**. It's a useful concept and when you divide the binding energy by the number of nucleons it's an excellent measure of the stability of an individual nucleus.

Example

For the $^4_2$He nucleus, calculate:

(a) the binding energy and

(b) the binding energy per nucleon.

Data: nuclear mass of $^4_2$He = 4.001506 u, mass of proton (m_p) = 1.007276 u, mass of neutron (m_n) = 1.008665 and 1 u ≡ 931 MeV

Answer

(a) First, you know that you have 2 protons (from the atomic number) and that you have 2 neutrons (mass number – atomic number).

Next, add up the individual masses of the particles:

$$2 \times m_p + 2 \times m_n = 2 \times 1.007276 + 2 \times 1.008665 = 4.031882 \text{ u}$$

This is the mass of the nucleons before they were brought together and the mass after they were brought together is 4.001506 (i.e. the mass of $^4_2$He). The difference is:

$$4.031882 - 4.001506 = 0.030376 \text{ u} \quad \text{which is the decrease in mass}$$

$$\text{Binding energy} = 0.030376 \times 931 = 28.28 \text{ MeV}$$

(b) All you have to do for the second answer is divide by the number of nucleons (i.e. divide by 4, because of the 2 neutrons and 2 protons)

$$\text{B.E / nucleon} = 28.28 / 4 = 7.07 \text{ MeV / nucleon}$$

Tricky synoptic example

A spring is extended 8.2 cm using a force of 360 N. Calculate the increase in mass of the spring.

Answer

This question may seem like science fiction nonsense but it's actually quite easy and real.

$$\text{PE} = \tfrac{1}{2}Fx = \tfrac{1}{2} \times 360 \times 8.2 \times 10^{-2} = 14.8 \text{ J}$$

$$E = 14.8 = mc^2 \quad \rightarrow \quad m = \frac{14.8}{c^2} = 1.6 \times 10^{-16} \text{ kg}$$

The increase in mass is because the atoms are being pulled farther apart as the spring is being extended and the **PE** is increasing.

≫ Pointer

Remember that binding energy is an energy given out that's associated with lost mass – **it's not an energy that a nucleus possesses** (alternatively it's an energy you have to provide in order to increase mass when the nucleons are pulled apart).

quicKfire

⑤ The mass of an atom of C12 is 12 u (exactly) by definition.

Calculate the binding energy per nucleon of C12.

Data: m_{H1} = 1.007 825 u

m_n = 1.008 665 u

quicKfire

⑥ Calculate the binding energy per nucleon of $^{56}_{26}$Fe.

Data: m_{Fe56} = 55.934 939 u

m_p = 1.007 276 u,

m_n = 1.008 665 u

m_e = 0.000 548 u

quicKfire

⑦ A rock of mass 250 kg is lifted through a height of 2.14 m.

(a) Show that the mass of the potential energy change of the Earth-rock system is about 6×10^{-14} kg.

(b) Explain why the mass of the Earth-rock system does not increase by this amount.

3.8.3 The binding energy per nucleon vs nucleon number graph

This is the graph that shows the stability of nuclei. It also tells you if nuclei are likely to perform fusion or fission reactions.

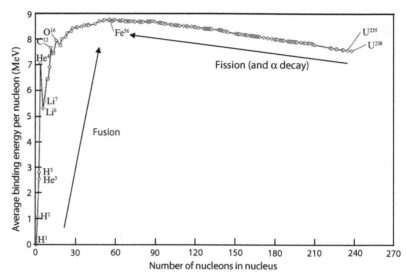

Fig. 3.8.2 Binding energy curve

You should notice that $^4_2$He and $^{56}_{26}$Fe are plotted correctly (if you obtained the correct answer to Quickfire 5). You should also notice that $^1_1$H has 0 binding energy per nucleon – that's quite obvious after you've thought about it. The $^1_1$H nucleus is just a proton which cannot have any binding energy because there's nothing else in the nucleus with it (see **Pointer**). Furthermore, $^{56}_{26}$Fe is close to the maximum of the curve and is one of the most stable of all nuclei.

All this means that smaller nuclei can undergo fusion to increase their nucleon number and move towards the stable part of the graph (see the fusion arrow on the graph). Heavy nuclei will undergo fission (or α decay) to decrease their nucleon number and move towards stability.

Trickier example

One of the fission reactions of $^{235}_{92}$U (uranium-235) is the following:

$$^{235}_{92}\text{U} + ^1_0\text{n} \quad \rightarrow \quad ^{95}_{37}\text{Rb} + ^{137}_{55}\text{Cs} + 4^1_0\text{n}$$

i.e. $^{235}_{92}$U is hit by a neutron and breaks up into $^{95}_{37}$Rb, $^{137}_{55}$Cs and 4 neutrons.

mass of $^{235}_{92}$U = 235.0439 u,   mass of $^{95}_{37}$Rb = 94.9293 u,

mass of $^{137}_{55}$Cs = 136.9071 u,   mass of neutron = 1.0073 u

Calculate the energy released in the reaction and explain your answer in terms of binding energy and stability.

> ### ≫ Pointer
>
> Actually, a $^1_1$H does have a binding energy. It is 13.6 eV, which is the energy needed to ionise a hydrogen atom. This is far too small to be seen on the scale of Fig. 3.8.2.

⑧ Explain briefly why lighter elements, in general, give off energy when they are involved in fusion reactions.

Answer

Total mass of LHS = 235.0439 + 1.0073 = 236.0512 u

Total mass of RHS = 94.9293 + 136.9071 + 4 × 1.0073 = 234.8656 u

mass lost = 0.1856 u

Finally, energy released = 0.1856 × 931 = 173 MeV

In terms of binding energy, the products $^{137}_{55}$Cs and $^{95}_{37}$Rb have fewer nucleons than $^{236}_{92}$U and their binding energy per nucleon is greater than $^{236}_{92}$U (i.e. followed the fission arrow on the graph). The products therefore are more stable, have a smaller total mass and an extra 173 MeV of energy is released.

If you really wanted to give a complete answer, you could also state that the binding energy of the free neutrons is zero and that they decay into a proton, electron and anti-neutrino with a half-life of 10 minutes (this would not be expected though).

Even Trickier Example

Use the binding energy per nucleon values from the Fig. 3.8.2 to estimate the energy released in the following reaction.

$$^2_1H + {}^3_1H \rightarrow {}^4_2He + {}^1_0n$$

Answer

No masses are provided so you must use the binding energy per nucleon values from the graph (as the question says). Looking carefully at the graph you should agree that approximately:

BE($^2_1$H) = 1.1 MeV/nucleon   BE($^4_2$He) = 7.1 MeV/nucleon

BE($^3_1$H) = 2.8 MeV/nucleon

You should also be able to state that the binding energy of the neutron is zero. Now, to calculate the total binding energy we must multiply the binding energies per nucleon with the nucleon number.

Total binding energy of LHS = 2 × 1.1 + 3 × 2.8 = 10.6 MeV

Total binding energy of RHS = 4 × 7.1 = 28.4 MeV

So the **increase** in binding energy is 17.8 MeV. From the definition of the binding energy, this is the energy that is released in the reaction. Alternatively, this **increase** in the binding energy of the nuclei is the same as the **decrease** in the total potential energy of the nuclei and so is the energy **released** in the reaction.

---

quickfire

⑨ Another possible fission reaction involving uranium-235 is shown. Calculate the energy released from the data provided.

$$^{235}_{92}U + {}^1_0n \rightarrow$$
$$^{89}_{36}Kr + {}^{144}_{56}Ba + 3{}^1_0n$$

mass of $^{235}_{92}$U = 235.0439 u,

mass of $^{89}_{36}$Kr = 88.9176 u,

mass of $^{144}_{56}$Ba = 143.9230 u,

mass of neutron = 1.0087 u

quickfire

⑩ Use the binding energy per nucleon values from the graph to estimate the energy released in the 'trickier example' and confirm that the value obtained in Quickfire 9 is correct.

>> **Pointer**
When binding energy is **gained**, energy is released

>> **Pointer**
When potential energy is **lost**, energy is released

>> **Pointer**
The energy released in nuclear reactions usually takes the form of kinetic energy of the products.

Data for use as appropriate in the questions below

Atomic mass data (in u): $^1_1$H(1.007 825); $^2_1$H(2.014 102); $^3_1$H(3.016 049); $^3_2$He(3.016 030); $^4_2$He(4.002 604); $^8_4$Be(8.005 305); $^{11}_5$B(11.009 305); $^{11}_6$C(11.011 434); $^{14}_6$C(14.003 241); $^{14}_7$N(14.003 074); $^{56}_{26}$Fe(55.934 937); $^{61}_{28}$Ni(60.931 056); $^{103}_{40}$Zr(102.926 601); $^{93}_{41}$Nb(92.906 378); $^{103}_{45}$Rh(102.905 504); $^{134}_{54}$Xe(133.905 395); $^{137}_{56}$Ba(136.905 827); $^{234}_{90}$Th(234.043 601); $^{233}_{92}$U(233.039 635); $^{238}_{92}$U(238.050 788); $^{239}_{94}$Pu(239.052 163); $^{244}_{95}$Am(244.064 285)

Particle masses (in u): $^0_{-1}$e(0.000 548); $^1_0$n(1.008 665); $^1_1$p(1.007 276); $^0_0\overline{v}_e$ (≈ 0);

Other data: 1 u = 1.661 × 10^{-27} kg ≡ 931 MeV; c = 3.00 × 10^8 m s^{-1}; m_e = 9.11 × 10^{-31} kg

1. (a) Convert all the following masses to the unified mass unit (u):
   (i) 2.3 kg   (ii) 42 mg   (iii) 9.11×10^{-31} kg   (iv) 6.0 tonne   (v) 11.2 fg
   (b) Convert the following masses to kg:
   (i) 56 u   (ii) 4×10^{30} u   (iii) 5.8 µg   (iv) 1.20×10^{57} u (what body has this mass?)

2. Use data from the box above to calculate the mass of a He4 **nucleus**. Give your answer (a) in u, (b) in kg.

3. Calculate the energies released in all the following reactions:
   (a) $^2_1H + {}^3_1H \rightarrow {}^4_2He + {}^1_0n$ (check your answer with the 'Even trickier example')
   (b) $^{14}_6C \rightarrow {}^{14}_7N + {}^0_{-1}e + {}^0_0\overline{\nu}_e$
   (c) $^{238}_{92}U \rightarrow {}^{234}_{90}Th + {}^4_2He$
   (d) $^4_2He + {}^4_2He \rightarrow {}^8_4Be$

4. Calculate the energy release when a 1 kg mixture of ^2_1H and ^3_1H in the correct ratio undergoes the reaction in 3(a). State the mass of ^2_1H required.

5. Calculate the binding energies per nucleon of the following nuclei:
   (a) $^{11}_5B$ (b) $^{61}_{28}Ni$ (c) $^{244}_{95}Am$ (d) ^8_4Be

6. (a) Use the binding energy per nucleon graph to explain why only heavy elements undergo alpha particle decay.
   (b) The reaction in question 3(d) is slightly endothermic (absorbs rather than releases energy). Explain whether or not the binding energy per nucleon you calculated in 5(d) should be greater or smaller than that of ^4_2He (7.07 MeV/nucleon).
   (c) Explain why the reaction $^3_1H \rightarrow {}^3_2He + {}^0_{-1}e + {}^0_0\overline{\nu}_e$ appears to contradict information in the binding energy per nucleon curve (Fig. 3.8.2).

7. Use Fig. 3.8.2 to estimate the energy released in the *triple alpha reaction*, $^4_2He + {}^4_2He + {}^4_2He \rightarrow {}^{12}_6C$ which occurs in the red giant stage of stars.

8. Fissile nuclei have approximately 240 nucleons. The fission products can be taken as having nucleon numbers around 120.
   (a) Use Fig. 3.8.2 to show that the energy released in a typical fission reaction is approximately 200 MeV.
   (b) A nuclear fission reactor has a constant thermal output of 1.5 GW. Use your answer to part (a) to estimate the mass of nuclear fuel that needs replacing each year.

9. Nuclear reactors produce a great range of fission products. One fission reaction of Pu239 is:
   $$^1_0n + {}^{239}_{94}Pu \rightarrow {}^{134}_{54}Xe + {}^{103}_{40}Zr + 3{}^1_0n$$
   Xe134 is stable but Zr103 decays in a series of steps to Rh103.
   (a) Identify the nature of the decay undergone by Zr103.
   (b) Calculate the energy released (in MeV) in the initial fission reaction.
   (c) Calculate the percentage of the total energy release which is due to the decay of the Zr103 to Rh103.

10. Uranium-233 is also fissile. A possible reaction is:
    $$^1_0n + {}^{233}_{92}U \rightarrow {}^{137}_{54}Xe + {}^{93}_{38}Sr + 4{}^1_0n$$
    Both the fission products are radioactive and decay in a series of steps to Ba137 and Nb93. Calculate the total energy released in the whole process.

3.9 Magnetic fields

Magnetic fields are also known as B-fields from the symbol, B, which is used for the strength of the magnetic field.

3.9.1 The force on a wire carrying current in a magnetic field

Have a look at the wire shown between the poles of a strong magnet. Wires carrying a current at an angle to a magnetic field will experience a force. The force is given by the equation

$$F = BIl\sin\theta$$

where B is the magnetic flux density (or B-field), I is the current, l is the length of the wire in the B-field and θ is the angle between the wire and the magnetic field.

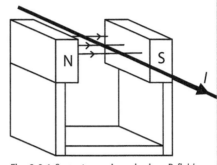

Fig. 3.9.1 Current-carrying wire in a B-field

To obtain the maximum force (for a given field, wire and current) you need $\sin\theta = 1$.

The angle θ should be 90° i.e. the wire should be at right angles to the magnetic field (B-field). In the 3D diagram shown, the wire passes through the B-field perpendicular to the field lines so you can simplify the equation (in this case) to

$$F = BIl$$

Now, let's calculate the greatest force you can exert on a wire in a typical lab. The maximum strength magnet you're likely to have in your lab will have $B = 0.2$ T (roughly) with a length between the poles of around 5 cm. The maximum current you're likely to get out of your most expensive power supply will be around 10 A. This gives a maximum force (using the above set-up) of:

$$F = 0.2 \times 10 \times 0.05 = 0.1\,\text{N}$$

Not very impressive really. Nonetheless, this is the effect that makes all electric motors work from the motor in your Blu-ray player to the starter motor that starts your car every morning.

What about the direction of the force? This is one of the trickiest things to do at A-level because you have to think in 3D. You need to use Fleming's left-hand rule (or FLHR for short).

Look at Fig 3.9.2. You need to align your **F**irst finger along the direction of the **F**ield (B-field) and your se**C**ond finger along the direction of the **C**urrent. Your thu**M**b will then point out the direction of **M**otion.

>> Pointer

Note that the force is always at right angles to both the field and the current.

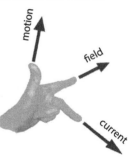

Fig. 3.9.2 Fleming's left-hand rule

quickᖴiᖇe

① State the direction of the forces on the wires

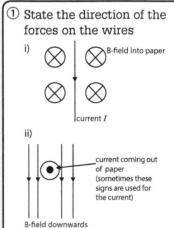

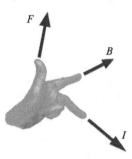

Fig. 3.9.4 FLHR (2)

If you apply this rule to the initial set-up above, you'll find you don't have to move your left hand much from that in the diagram on the right to find that your thumb is pointing upwards indicating that the force on the wire is upwards. In exam papers, the set-up won't be quite so simple and you may have to contort your left hand into a strange position to find the correct direction. Sometimes the magnetic field will not be represented by field lines but rather by arrow heads or arrow tails:

(•) Signifies a field (or a current) coming out of the paper

Signifies a field (or a current) going into the paper

⊗ Fig. 3.9.3 Direction indication

Examples

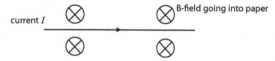

Fig. 3.9.5 FLHR example 1

Using FLHR, you should be able to find that the force on the above wire is up.

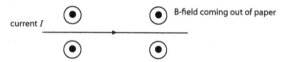

Fig. 3.9.6 FLHR example 2

Now, if you turn your left hand upside down and point your first finger towards your face you should find that the force on the wire in the second example is down.

3.9.2 Force on a charge moving in a magnetic field

This theory is very similar to the force on a wire in a magnetic field and you'll need to use FLHR again to predict the direction of the force. The equation you'll be using is:

$$F = Bqv\sin\theta$$

where B, once again, is the magnetic flux density (or B-field), q is the size of the moving charge, v is the velocity of the moving charge and θ the angle between the velocity and the B-field. You can actually derive the equation $F = BIl\sin\theta$ using $F = Bqv\sin\theta$ and $I = nAve$.

Example

A proton in a magnetic field. Calculate the magnitude and direction of the force on the proton.

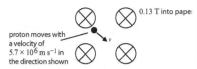

Fig. 3.9.7 Proton in a magnetic field

Answer

Perhaps this might be difficult to picture at first (you need to think in 3D) but the proton is moving at right angles to the B-field. Hence $\theta = 90°$, and

$$F = Bqv\sin 90° = Bqv = 0.13 \times 1.60 \times 10^{-19} \times 5.7 \times 10^6 = 1.2 \times 10^{-13}\,\text{N}$$

To find the direction, you use FLHR and use the direction of the proton velocity as the direction of the current. After pointing your first finger into the paper and rotating you should find that the direction of the force is as shown (Fig 3.9.8). Note that the force is at right angles to the velocity (and to the B-field).

Fig. 3.9.8 Direction of force on proton

Interestingly, if you consider a proton with the same speed but a slightly different direction (see Fig 3.9.9), the force will be the same magnitude but with a slightly different direction. With a net force at right angles to the motion you should see that this set-up will give circular motion. What is the radius of motion of the proton?

Fig. 3.9.9 Changing direction

You should remember from circular motion in Section 1.5

$$F = \frac{mv^2}{r}$$ and you can equate this to the magnetic force $F = Bqv$

so $\quad \dfrac{mv^2}{r} = Bqv \;\rightarrow\; \dfrac{mv}{r} = Bq \;\rightarrow\; r = \dfrac{mv}{Bq}$

Putting in the actual numbers

$$r = \frac{1.67 \times 10^{-27} \times 5.7 \times 10^6}{0.13 \times 1.60 \times 10^{-19}} = 0.46\,\text{m (or 46\,cm)}$$

The fact that charged particles tend to perform circular motion in magnetic fields is extremely useful and led to the development of the TV, the discovery of the mass of the electron as well as being used in particle accelerators and mass spectrometers.

Example

The diagram shows an electron moving in a circular path in a B-field that is coming out of the paper. Place an arrow on the path to show the direction of motion of the electron.

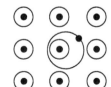

Fig. 3.9.10 Electron in a B-field

Answer

You need to apply FLHR to some point on the electron's path and remember that the force (thumb) is towards the centre of the circle (it's a centripetal force). Your second finger will then give you the current. Unfortunately, the charge on an electron is negative, so you have to reverse the direction of the current to get the direction of motion of the electron. You should eventually arrive at the answer that the electron is going anticlockwise along the circular path.

quickfire

④ Derive the equation $F = BIl\sin\theta$ by considering a force of $F = Bqv\sin\theta$ on electrons in a wire of cross-sectional area A and length l in a uniform field B.

quickfire

⑤ Calculate the force on an electron travelling at a velocity of 25×10^6 m s^{-1} at an angle of 35° to a uniform B-field of 3.4×10^{-3} T.

Grade boost

Remember how to derive $r = \dfrac{mv}{Bq}$ using $\dfrac{mv^2}{r} = Bqv$. There are often questions on circular motion due to charged particles in B-fields.

quickfire

⑥ By equating the other expression for the centripetal force ($m\omega^2 r$) to the force on a charged particle (Bqv), derive the expression $\omega = \dfrac{Bq}{m}$ (which is the equation that led to the design of the cyclotron).

Grade boost

You'll often have to apply FLHR, use $Bev = Ee$, $E = \dfrac{V}{d}$ and even calculate n (giving its unit as m^{-3}).

quickfire

⑦ Show how you'd connect a voltmeter to the Hall probe to measure the Hall voltage.

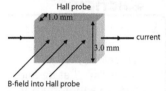

Hall probe
1.0 mm
3.0 mm
current
B-field into Hall probe

quickfire

⑧ The Hall voltage in Quickfire 7 is 3.6×10^{-6} V. Calculate the electric field associated with this Hall voltage.

Grade boost

If you're a B–A* candidate, you should be comfortable in deriving all the Hall effect equations – just remember the principles and the equations will come naturally. If you're not comfortable deriving the equations memorise them – some weaker candidates earn extra marks with this tactic.

3.9.3 The Hall probe

This is a particularly important device for measuring B-fields but it is also used continuously in research and high-tech facilities to measure electron properties in semiconductor chips. Here's how it works:

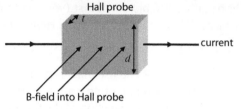

Hall probe
t
current
d
B-field into Hall probe

Fig. 3.9.11 The Hall probe

If you apply FLHR to the set-up, you should find that the force is upwards on the free electrons that are providing the current. This means that the free electrons will move toward the top face of the Hall probe making the top of the Hall probe negatively charged. This cannot carry on forever because the electrons will be repelled by the negative charge on the top surface. Very soon, an equilibrium will be reached when the magnetic force (Bqv or Bev) is balanced by the electric repulsion force. But what is this electric repulsion force? You should remember that the force is the electric field multiplied by the charge (Eq or Ee from Section 2.6). Now for a little bit of algebra:

Magnetic force = electric force

$$Bev = Ee \quad \rightarrow \quad Bv = E$$

but the electric field can be related to the pd between the bottom and top plate using the equation $E = \dfrac{V}{d}$ (same as for a capacitor). Hence,

$$Bv = \frac{V}{d} \quad \rightarrow \quad V = Bvd$$

and it's this pd across the Hall probe that you can actually measure using a simple voltmeter. This pd is usually called the Hall voltage (V_{H}). You get an idea of how this Hall voltage is useful when you realise that v is the drift velocity of the electrons. So you can measure the drift velocity if you know the dimensions of the Hall probe (in this case d) and the B-field (V_{H} you can measure with your voltmeter).

$$v = \frac{V_{\mathrm{H}}}{Bd}$$

You can go even further if you've got an ammeter and know the equation $I = nAve$.

Rearranging for v gives $v = \dfrac{I}{nAe}$ then substitute in $v = \dfrac{V_{\mathrm{H}}}{Bd}$

$$\frac{I}{nAe} = \frac{V_{\mathrm{H}}}{Bd} \quad \rightarrow \quad n = \frac{IBd}{V_{\mathrm{H}}Ae}$$

You can simplify this expression for n further when you realise that the cross-sectional area (A) of the probe is $A = t \times d$:

$$n = \frac{IBd}{V_H t \times de} = \frac{IB}{V_H te}$$

Quite incredible when you think about it. You can actually measure the drift velocity and number of free electrons per m^3 just by using a cheap little ammeter and voltmeter (although you do need to know your B-field and d and t for your probe).

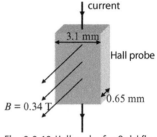

≫ Pointer

You can rearrange

$$n = \frac{IB}{V_H te} \text{ to get } V_H = \frac{BI}{nte}$$

i.e. the Hall voltage is proportional to the B-field. This is why Hall probes are often used to measure B.

(a) Tricky Hall probe

Example

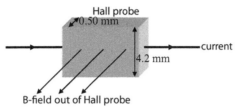

Fig. 3.9.12 Hall probe example

(i) Connect a voltmeter correctly to the probe to show how you would measure the Hall voltage.

(ii) Explain which face of the probe will become positive (the current is due to free electrons).

(iii) The Hall voltage is $820\,nV$ and the B-field is $0.14\,T$. Calculate the drift velocity of the free electrons.

(iv) The current is $0.47\,mA$. Calculate the number of free electrons per unit volume.

Fig. 3.9.13 Hall probe for Quickfire 9

Answer

(i) Attach the voltmeter (right) to the top and bottom face of the above Hall probe.

(ii) The force is down (using FLHR) so the bottom face becomes negative and the top face becomes positive (due to a deficiency of electrons).

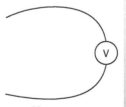

(iii) From $Bev = Ee \rightarrow Bv = E \rightarrow Bv = \frac{V_H}{d} \rightarrow v = \frac{V_H}{Bd}$

so $v = \frac{V_H}{Bd} = \frac{820 \times 10^{-9}}{0.14 \times 4.2 \times 10^{-3}} = 1.39 \times 10^{-3}\,m\,s^{-1}$ (or $1.39\,mm\,s^{-1}$)

(iv) The equation has been derived earlier but here it is again

$$\frac{I}{nAe} = \frac{V_H}{Bd} \rightarrow n = \frac{IBd}{V_H Ae} = \frac{IB}{V_H te}$$

$$= \frac{0.47 \times 10^{-3} \times 0.14}{820 \times 10^{-9} \times 0.50 \times 10^{-3} \times 1.6 \times 10^{-19}}$$

$$n = 1.00 \times 10^{24}\,m^{-3}$$

quickfire

⑨ a) In Fig. 3.9.13, show how you would connect a voltmeter to the Hall probe to measure the Hall voltage.

b) The charge carriers are free electrons. Explain why the right face of the probe becomes positive.

c) The Hall voltage is $1.35\,\mu V$. Calculate:
   i) the Hall field (E)
   ii) the drift velocity of electrons.

d) The current is $870\,\mu A$. Calculate n and give its unit.

(b) Using a Hall probe to measure a B-field

This is the last thing you need to know about the Hall probe.

When measuring a B-field with a Hall probe you simply need to:

(i) place the probe in the field and

(ii) orientate it so that its front face is at right angles to the B-field (alternatively you can just twiddle it around until you get a maximum reading).

>> **Pointer**
You need to learn these sketches of field lines but you may remember them from GCSE.

quickfire

⑩ Sketch the field lines

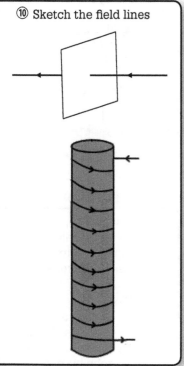

3.9.4 The magnetic field

Here are two magnetic fields that you should be able to sketch.

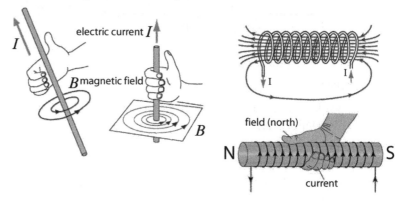

Fig. 3.9.14 Magnetic field patterns

The two diagrams on the left in Fig. 3.9.14 show the magnetic field lines due to a long wire and the two diagrams on the right show the magnetic field due to a long solenoid (cylindrical coil of wire). You'll need to know the directions of the field also but the good news is that the right-hand grip rule works for both (see diagrams). For the straight wire, place your thumb in the direction of the current and your grip follows the direction of the field lines. For the solenoid, let your grip follow the route of the current and your thumb will point in the direction of the B-field **inside** the solenoid.

You'll also need to do calculations based on the magnetic field of a long wire and the magnetic field of a long solenoid (both equations are in the Data booklet).

For a long wire, the magnetic flux density (B-field) is given by

$$B = \frac{\mu_0 I}{2\pi a}$$

where μ_0 is the permeability of free space, I is the current in the wire and a is the shortest distance to the wire.

Inside a long solenoid, the magnetic flux density is given by

$$B = \mu_0 n I$$

>> **Pointer**
The permeability of free space, μ_0, has a defined value of $4\pi \times 10^{-7}$ H m^{-1}.

where μ_0 is still the permeability of free space, I is the current in the solenoid and n is the number of turns per unit length of the solenoid.

Examples

(i)   Calculate the magnetic flux density at a distance of 2.7 mm from a long wire carrying a current of 5.2 A.

(ii) A solenoid of length 145 cm has 26 500 turns and carries a current of 358 mA. Calculate the magnetic flux density inside the solenoid.

Answers

(i) $B = \dfrac{\mu_0 I}{2\pi a} = \dfrac{4\pi \times 10^{-7} \times 5.2}{2\pi \times 2.7 \times 10^{-3}} = 0.39$ mT

(ii) $n = \dfrac{26\,500}{1.45} = 18\,300$ turns m^{-1}

$B = 4\pi \times 10^{-7} \times 18\,300 \times 0.358 = 8.23$ mT

One final thing that you need to know about solenoids is this. An iron core inside a solenoid increases the magnetic flux density greatly. From the equation $B = \mu_0 n I$ you should see that doubling the current will double the magnetic flux density and likewise the current. However, putting a nearly pure iron core will increase the B-field by a factor of many thousands (even up to 200 000 times for pure iron).

3.9.5 Force between two wires carrying a current

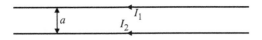

Fig. 3.9.15 Two current-carrying wires

When two wires carry a current they will exert forces on one another. The reason for this is as follows:

1 The top wire has a magnetic field.

2 The bottom wire is in this field.

3 The bottom wire 'feels' a force due to $F = BIl\sin\theta$.

Of course, the reverse argument is absolutely valid:

1 The bottom wire has a magnetic field.

2 The top wire is in this field.

3 The top wire 'feels' a force due to $F = BIl\sin\theta$.

How large is the force and what's its direction?

To answer this you need the formula

$$B = \frac{\mu_0 I_1}{2\pi a}$$

quickpire

⑪ A solenoid has a B-field inside it of 1.75 mT while carrying a current of 0.240 A. Calculate the number of turns per metre of the solenoid.

quickpire

⑫ A long wire carries a large constant current of 214 A. Where is the magnetic flux density greater than 1 mT?

quickpire

⑬ Use FLHR and the right-hand grip rule to show that the magnetic force acting between the two parallel wires is repulsive.

quickpire

⑭ Calculate the force per unit length acting on the long wires shown.

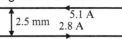

quickfire

⑮ Show that the force acting on the middle wire is zero (hint: the middle wire is in the field due to the other two wires, so you only need to consider the field due to the top and bottom wires).

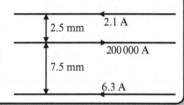

2.5 mm 2.1 A

200 000 A

7.5 mm

6.3 A

Key Term

Electron volt = the energy transferred when an electron moves between two points with a potential difference of 1 volt between them. So, for an electron being accelerated in a vacuum, it is the KE acquired when accelerated through a pd of 1V.

$(1 \text{ eV} = 1.6 \times 10^{-19} \text{ J})$

≫ Pointer

The constant acceleration of the electrons means that you can apply the constant acceleration equations to the electron motion.

Grade boost

Some students remember the equation

$F = Eq$

by calling it the Father Ted equation! (You might have to research Father Ted Crilly to understand why.)

for the field due to the top wire in the position of the bottom wire. You should recall the shape of the circular field lines around the wire and realise that the B-field (due to the top wire) will be coming out of the paper at the position of the bottom wire. This is at right angles to the direction of the current in the wire, so $\sin\theta = 1$.

Now using $F = BIl\sin\theta$ for the bottom wire.

$$F = BIl\sin\theta = \frac{\mu_0 I_1}{2\pi a} \times I_2 \times l\sin\theta = \frac{\mu_0 I_1 I_2 l}{2\pi a}$$

For the direction you'll need FLHR. The field is coming out of the paper so point your first finger towards your face and you should find that the force on the bottom wire is upward. By Newton's 3rd law, the force on the top wire must be of equal size and downward (i.e. two parallel wires carrying a current in the same direction experience an attractive force).

Let's see if it's possible to get a big force between wires in a lab. You might be able to get your hands on the 10 A power supply again and have two 5 m long wires maybe 1 mm apart. This gives a force of

$$F = \frac{\mu_0 I_1 I_2 l}{2\pi a} = \frac{4\pi \times 10^{-7} \times 10 \times 10 \times 5}{2\pi \times 0.001} = 0.1 \text{ N} \quad \text{(again, not a large force)}$$

3.9.6 Ion beams and accelerators

You also need to know a little about the effect of magnetic and electric fields on ion beams and this leads naturally to particle accelerators. You don't have to memorise the details of the particle accelerators but you need to be able to apply the physics you already know to them.

Let's start with the very first particle accelerator that was invented in the 1850s. This was just an empty glass tube with a cathode and anode to accelerate electrons.

Fig. 3.9.16 Early particle accelerator

Similar to a capacitor, there's a uniform electric field between the cathode and the anode. This is what accelerates the electrons and you can calculate the acceleration using the following equations (note that the first equation, which defines the electric field, is not on the Data booklet).

$$F = Eq$$

You can combine the above equation with $E = \frac{V}{d}$ and $F = ma$ to give

$$a = \frac{Vq}{md}$$

If the distance between the cathode and the anode is 10 cm and you apply a pd of 100 V. You get rather a large acceleration:

$$a = \frac{Vq}{md} = \frac{100 \times 1.6 \times 10^{-19}}{9.11 \times 10^{-31} \times 0.1} = 1.8 \times 10^{14} \text{ m s}^{-2}$$

How much energy has the electron gained just before it reaches the anode?

Using
$$W = q\Delta V_E$$
$$= 1.6 \times 10^{-19} \times 100 = 1.6 \times 10^{-17} \text{ J}$$

This is where a new unit of energy is defined – the electron-volt (eV). It's the energy gained by an electron when accelerated through a pd of 1V i.e. the energy gained by the electron above was 100 eV because it was accelerated through 100 V.

The particle accelerator in Fig 3.9.17 is slightly more complicated. It also has a vertical electric field to deflect the electrons. Obviously the electrons will be attracted to the positively charged plates.

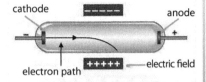

Fig. 3.9.17 Particle accelerator Mk II

This means that they are accelerated from left to right but are also deflected downwards as shown. The upper and lower plates will behave like a capacitor and there will be a uniform field between them. The electrons therefore will experience a constant force downwards of $qE_{vertical}$.

(a) The linear accelerator (Linac)

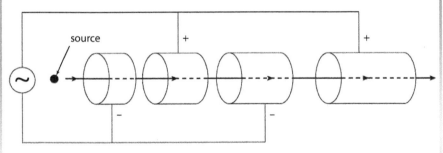

Fig. 3.9.18 Linear accelerator – Linac

This is a series of tubes that are charged either +ve or −ve depending on the alternating pd sent to them. First, let's say that a proton is accelerated to the right towards the −ve tube. When the proton arrives inside the tube there is no force acting on it and this is when the direction of the pd changes. When the proton is in the next gap between the tubes, the second tube is now −ve and the electric field accelerates it to the right again. The important thing is to ensure that the pd is synchronised so that the proton is inside a tube as the pd changes direction. This is achieved by keeping the frequency constant but increasing the lengths of the tubes and the gaps between them (because the proton is travelling a greater and greater distance in the same time).

quickfire

⑯ Use $F = Eq$, $E = \dfrac{V}{d}$ and $F = ma$ to derive $a = \dfrac{Vq}{md}$

quickfire

⑰ Equate the 100 eV of energy for an electron to calculate the speed of an electron after it has been accelerated through 100 V.

quickfire

⑱ An electron is accelerated through a pd of 5.78 V. Calculate its final KE in
i) eV   ii) J

quickfire

⑲ A proton has a KE of 5.7 keV and it was accelerated from rest using an electric field.
i)   What was the pd with which the proton was accelerated?
ii) Calculate the KE of the proton in J.

>> **Pointer**
The electric field always increases the speed of the charged particles while the magnetic field keeps their paths circular.

quickfire

⑳ The alternating pd of a Linac is 125 kV. Through how many tubes must a proton pass before it has 750 keV of energy.

quickfire

㉑ In Quickfire 20 how would your answer change for
  i) an electron?
  ii) a helium nucleus (hint: with charge +2e)?

quickfire

㉒ Why can't particle accelerators accelerate neutrons?

quickfire

㉓ Calculate the frequency of a cyclotron if it's accelerating electrons in a field of 0.115 T.

(b) The cyclotron

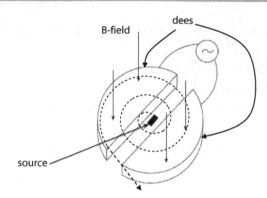

Fig. 3.9.19 The cyclotron

Again, the acceleration (or speed increase) is provided by an electric field. When a proton (say) is in the gap between the two Dees (semi-circular plates) it's accelerated across the gap by an electric field. The magnetic field shown keeps the proton in a circular motion but as the speed of the proton increases so does the radius of its circle. Hence the proton spirals out and eventually leaves the cyclotron.

This is where the answer to Quickfire 6 is handy. You can calculate the frequency of the pd from the theory:

$m\omega^2 r = Bqv$ but from circular motion $v = \omega r$

$$m\omega^2 r = Bq\omega r$$

and dividing by ωr and remembering that $\omega = 2\pi f$

$$\rightarrow m\omega = Bq \quad \text{leading to} \quad f = \frac{\omega}{2\pi} = \frac{Bq}{2\pi m}$$

Note that the frequency is a constant because the B-field is uniform and q and m are both constants. This is the beauty of the cyclotron – the frequency stays the same even as the velocity of the charged particle is increasing.

Example

Calculate the frequency of the pd supply for a cyclotron accelerating protons in a uniform magnetic flux density of 4.22 T, $(m_p = 1.67 \times 10^{-27}\,\text{kg})$.

Answer

$$f = \frac{4.22 \times 1.60 \times 10^{-19}}{2\pi \times 1.67 \times 10^{-27}} = 64.3 \times 10^6\,\text{Hz (or 64.3 MHz)}$$

(c) The synchrotron

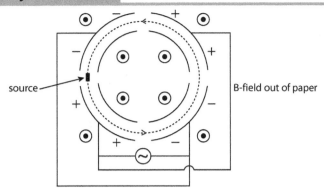

Fig. 3.9.20 The synchrotron (simplified)

Fig 3.9.20 is a simplified synchrotron but it shows the fundamental principles of its operation. The speed increase is again provided by the alternating pd and, again, the charged particles will be performing circular motion due to the B-field. The acceleration now occurs four times per 'orbit' when the particles are crossing between the differently charged tubes.

In contrast to the cyclotron, however, the path remains constant (same radius) so the B-field must increase in strength as the particles move more quickly. Also, the frequency of the ac supply must increase as the particles' speed increases.

Example

A synchrotron operates with a pd of $30\,\text{kV}$. What is the increase of KE of a helium nucleus after it has completed 8 cycles of the synchrotron?

Answer

A helium nucleus has a charge of +2e, therefore each time it crosses a gap it gains 60 keV of energy. Each cycle means being accelerated four times, hence the increased KE will be:

$$60\text{ keV} \times 4 \times 8 = 1920\text{ keV} (= 1.92\,\text{MeV})$$

Example

Calculate the speed of a helium nucleus with a KE of 1.92 MeV.

Answer

First the conversion $\qquad 1.92\text{ MeV} = 1.92 \times 10^6 \times 1.6 \times 10^{-19}\text{ J}$

then rearranging $\mathbf{KE} = \frac{1}{2}mv^2$ gives $v = \sqrt{\dfrac{2 \times \text{KE}}{m}}$

A mass of $4\,\textbf{u}$ is good enough for this calculation but this would be given in an exam question

$$v = \sqrt{\dfrac{2 \times 1.92 \times 10^6 \times 1.6 \times 10^{-19}}{4 \times 1.66 \times 10^{-27}}} = 9.62 \times 10^6\text{ m s}^{-1}$$

---

quicKfire

㉔ In the diagram of the synchrotron, use FLHR to work out whether the particle being accelerated is +ve or -ve.

≫ *Pointer*

You often have to work out speeds of particles with a particular KE, $v = \sqrt{\dfrac{2 \times KE}{m}}$ is quite a useful equation to remember.

quicKfire

㉕ The speed of an electron being accelerated in a synchrotron is doubled. What is the change in:
  i)   the frequency of the a.c. voltage?
  ii) the B-field?
  iii) the KE of the electron?

Example

The radius of a synchrotron is $4.80\,\text{m}$. Calculate the (instantaneous) magnetic flux density (B) and frequency of the pd when the particles being accelerated are helium nuclei with a speed of $9.62 \times 10^6\,\text{m s}^{-1}$.

Answer

Using $\quad \dfrac{mv^2}{r} = Bqv \rightarrow B = \dfrac{mv}{qr} = \dfrac{4 \times 1.66 \times 10^{-27} \times 9.62 \times 10^6}{2 \times 1.6 \times 10^{-19} \times 4.80} = 0.0416\,\text{T}$

and $\quad m\omega^2 r = Bq\omega r \quad f = \dfrac{Bq}{2\pi m} = \dfrac{0.0416 \times 2 \times 1.6 \times 10^{-19}}{2\pi \times 4 \times 1.66 \times 10^{-27}} = 320\,\text{kHz}$

However, this is the frequency of the circular motion. The frequency of the a.c. supply will be twice this (because in one period of the alternating supply the particle is accelerated twice and the particle is accelerated 4 times in one orbit of the synchrotron)

$$\text{Frequency of the a.c. supply} = 640\,\text{kHz}$$

>> **Pointer**
To convert a mass reading in gram to a force, you must first divide by 1000 to convert to kg and then multiply by ($9.81\,\text{N kg}^{-1}$).

3.9.7 Specified practical work

(a) Investigation of the force on a current in a magnetic field.

The easiest way to carry out this experiment accurately is to use a digital balance accurate to 0.01g. The apparatus required is shown in Fig. 3.9.21. There are many experiments that can be carried out based on the equation

$$F = BIl\sin\theta$$

the most obvious being to keep B, l and $\sin\theta$ all constant and investigate the relationship between the force and the current.

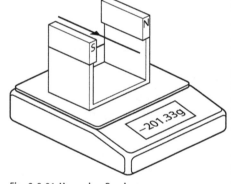

Fig. 3.9.21 Measuring B using an electronic balance

quickfire

㉖ Explain why the reading on the mass scale in Fig. 3.9.21 is negative.

In the diagram, the direction of the current is perpendicular to the magnetic field so that

$$F = BIl\sin\theta = Bl\,I$$

The equation has been put in the form $y = mx + c$ or more specifically, $y = mx$. So it should be clear that the force is proportional to the current and that the gradient of a F v I graph will be Bl. Some school physics departments will have expensive power supplies that can vary the current. These would make this experiment particularly easy to carry out. However, 6 D-type batteries combined with a current-limiting resistor of around 5 Ω should do the trick nicely. Your method would be something along the lines of:

- Place the U-shaped magnet on the mass scale.

- Press the tare button (to zero)

- Set up the wire so that it is well secured (won't move) and passes between the poles of the magnet as shown.

- Set up a series circuit to pass current through the wire using 1 D-type battery, a 5 Ω resistor, an ammeter and a switch.

- Close the switch and quickly measure the current on the ammeter and the mass reading on the scales.

- Repeat adding one more D-type battery up to a maximum of (around) six.

- Repeat the whole experiment again to obtain repeat readings.

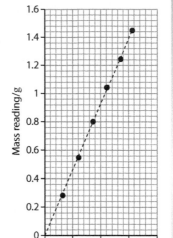

Fig. 3.9.22 Results graph for measuring B

This is a particularly accurate experiment if done properly and your results should be similar to Fig. 3.9.22 Note that the student has not converted the mass scale reading to a force so this will need to be done if obtaining a value of B (or l or Bl) from the gradient. Error bars should be drawn corresponding to ±0.01g on the y-axis and error bars corresponding to ±0.01A should be plotted on the x-axis (nearly all school ammeters will have this scale limited error for these currents).

quickfire

㉗ In Fig. 3.9.22, calculate:
   a) the gradient of the graph
   b) the magnetic flux density given that the length of the wire is 6.0 cm.

» Pointer

This set-up can also be used to investigate the relationship between the force and the length of the wire. However, you will need a set of wires of standard lengths (usually on a printed circuit board). A far cheaper DIY version is to use copper stripboard which has a standard hole separation of 2.54 mm. Hence, standard length wires of 5.08 mm, 10.16 mm, etc., can be obtained cheaply with a bit of soldering skill.

(b) Investigation of Magnetic Flux Density Using a Hall Probe.

Surprisingly enough, this experiment can be carried out even if your physics department does not have a Hall probe. There is a Hall probe in most smart phones nowadays so you can use the extremely accurate magnetic field detector in your smart phone (it is usually located in the top right corner of the phone). If you intend to use your smart phone you will probably need to download a suitable app but the good news is that there are many excellent free apps available.

The easiest investigation to carry out would be to investigate the relationship

$$B = \frac{\mu_0 I}{2\pi a}$$

for the magnetic field due to a long wire.

>> **Pointer**

Referring to the experiment of Fig. 3.9.23, if your smart phone provides a 3D vector of the B-field, you will probably need the z-component of the B-field.

If you are using a school Hall probe, ensure that the probe is flat on the desk so that the vertical component of the B-field is being measured.

quicKfire

㉘ Another investigation into the relationship

$$B = \frac{\mu_0 I}{2\pi a}$$

keeps the distance constant but varies the current. What graph would you plot and what gradient would you expect?

>> **Pointer**

Another possible use of a Hall probe would be to investigate the B-field inside a solenoid.

Investigation 1
Place the Hall probe in the middle of the solenoid and vary the current in regular steps, e.g. 0–3.0A in steps of 0.5A. Ideally the solenoid would be long and thin so that the equation $B = \mu_0 n I$ applies but a length 5× the diameter would be fine.

Investigation 2
As 1 but move the probe along the axis and investigate the variation of B with position.

Note: you will need a small Hall probe for these investigations as your smart phone probably won't fit inside the solenoid!

You should use the following set-up with the wire flat on a lab bench. You will need a large current (around 5 A) flowing in a long wire (1m should be enough). Your method will be along these lines:

1. Place the Hall probe/phone 1.0 cm from the long wire ($a = 1.0$ cm).

2. Switch on the current, record the Hall probe and current reading and switch off the current.

3. Increase the distance between the Hall probe/phone and the wire by 1.0 cm and repeat step 2.

4. Continue until a distance of 5.0 cm(ish) from the wire.

5. Repeat the whole experiment with the phone on the opposite side of the wire.

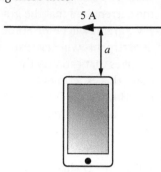

Fig. 3.9.23 Investigating B using a smart phone

In your analysis, you should subtract the B-field values obtained on both sides of the wire and divide by 2. This will eliminate the effect of the Earth's magnetic field and leave you with the value for the current's B-field (the B-field changes direction on either side of the wire). Also, because the exact location of the Hall sensor might not be known, it is best to rearrange the equation into the form

$$a = \frac{\mu_0 I}{2\pi} \times \frac{1}{B}$$

The actual distance to the Hall sensor will be $a = x + d$ where d is the distance from the top of the probe/phone to the Hall sensor. This then gives

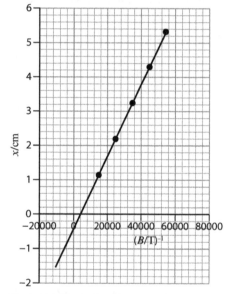

Fig. 3.9.24 Results from the smart phone investigation

$$x = \frac{\mu_0 I}{2\pi} \times \frac{1}{B} - d$$

A graph of x versus B^{-1} will be a straight line with gradient $\dfrac{\mu_0 I}{2\pi}$ and intercept $-d$.

Hence, this experiment can: (i) confirm the relationship $B = \dfrac{\mu_0 I}{2\pi a}$ for a long wire; (ii) obtain a reasonable value for μ_0 the permeability of free space and (iii) locate the Hall sensor in your smart phone.

x / cm	B_1 / µT	B_2 / µT	$\dfrac{B_1 - B_2}{2}$ /µT	$(B/\text{T})^{-1}$
1.0	123.8	−9.5	66.7	15 000
2.0	97.4	16.3	40.6	24 700
3.0	85.8	28.3	28.8	34 800
4.0	79.7	34.7	22.5	44 400
5.0	74.4	38.8	17.8	56 200

Table 3.9.1 Relationship between B and a – specimen results

» Pointer
The data plotted in Fig. 3.9.24 are based on the data in Table 3.9.1. Note how the Earth's magnetic field (around 56µT) has been eliminated by subtracting and halving columns 2 & 3.

1. State the direction of the force on the wire or charge carrier in the following diagrams.

(a) (b) (c) wire (d)

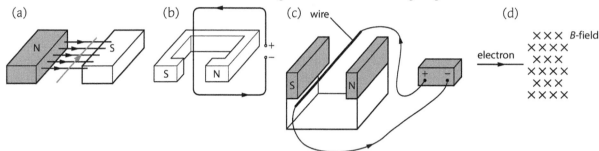

2. A charged particle in a uniform magnetic field in a vacuum will perform circular motion at constant speed.
   (a) Explain why the word 'velocity' was not used in the last sentence instead of speed.
   (b) By equating the magnetic force on a moving charge carrier to the centripetal force, explain (algebraically) why the frequency of performing circles is independent of the speed of the particle.
   (c) An ion performing circles in a uniform magnetic field of 125 mT has a charge of +e. If the frequency of circular motion is 636 kHz, calculate its mass in u.
   Chemists: identify the ion.
   (d) If the radius of the circular path of the ion is 12.5 m. Calculate:
   (i) the speed of the ion
   (ii) the KE of the ion in eV.

3. Suppose the charge carriers Hall probe in Fig. 3.9.11 are positive (known as 'holes'). The B-field is 0.240 T.
   (a) State which face of the Hall probe becomes +ve and which rule you used to obtain your answer.
   (b) The Hall voltage is measured as 1.29 mV and the width of the Hall 'chip' is 5.2 mm. Calculate the drift velocity of the holes.
   (c) The thickness of the 'chip' is 0.90 mm and the current is 0.57 A. Calculate the number of charge carriers per m³.
   (d) In light of your answers to (b) and (c) and the fact that the charge carriers are 'holes', what type of material is the 'chip' made of?

4. (a) A $_2^4$He nucleus is accelerated from rest in a LINAC. The pd between the tubes is $15\,kV$.

   (i) Calculate the change in energy of the helium nucleus between the 3rd and 4th tubes.

   (ii) Calculate the change in speed of the helium nucleus between the 3rd and 4th tubes.

   (iii) Calculate the frequency of the pd applied if the distance between the 3rd and 4th tubes is $2.3\,m$.

   (b) Positrons are accelerated in a cyclotron.

   (i) Calculate the magnetic flux density if the frequency of the pd is $120\,MHz$.

   (ii) The pd applied to the dees is $550\,V$. Calculate the energy of the positrons after $0.3\,\mu s$ if accelerated from rest.

   (iii) The maximum radius of the cyclotron is $18.0\,cm$. Calculate the speed of the positrons when they exit.

   (c) (i) Explain how the radius of accelerated particles remains constant in a synchrotron as their speeds increase.

   (ii) If a proton is to receive $1.2\,TeV$ by accelerating in a synchrotron and the accelerating pd of the synchrotron is $1.0\,MeV$. How many loops of the synchrotron must the proton complete?

   (iii) With the aid of a simple calculation, explain why Newtonian mechanics is inapplicable to a $1.2\,TeV$ proton.

5. The set-up of Fig. 3.9.21 is used to perform an investigation in which the length of the wire is varied from $1.0\,cm$ to $5.0\,cm$ but the magnetic field and the force are kept constant. The force on the wire is kept constant by adjusting the current so that the reading on the mass balance is $(0.50 \pm 0.01)\,g$. This experiment yields the results shown in the table.

Length l / cm (± 0.1 cm)	current, I / A (±2%)	$(I / A)^{-1}$ ±2%
1.0	4.21	
2.0	2.02	
3.0	1.39	
4.0	1.01	
5.0	0.80	

   (a) Complete the 3rd column, writing your answers to the correct number of significant figures.

   (b) Use a graph of 1/current v length of wire (with error) bars to determine the gradients of the steepest and least steep lines consistent with the error bars.

   (c) Use your answer to (b) to calculate the mean gradient along with its percentage uncertainty and hence determine the magnetic flux density and its uncertainty (remember: the mass reading is 0.50 g).

   (d) The manufacturer of the U-shaped magnet states that its flux density is $(125 \pm 5)\,mT$. Comment on the outcome of this investigation.

6. Use a graph of the data in Table 3.9.1, to obtain values for μ_0 and d (the distance from the top of the phone to the Hall sensor) together with their uncertainties, given that the current in the wire is $5.00\,A$. Assume an uncertainty in B^{-1} of 10% and in x of $\pm 1\,mm$.

3.10 Electromagnetic induction

This is the effect responsible for producing electricity and could well be the single most valuable discovery leading to our high standards of living in the 21st century. The bad news is that the concept of electromagnetic induction is quite difficult to understand and usually provides the lowest mark of all questions on A level exam papers! Moving wires and coils in magnetic fields (or stationary coils and wires in changing magnetic fields) all seem a little invisible and mysterious but it nearly all comes down to understanding one simple law which you'll come across later – Faraday's law.

Key Term

Magnetic flux:

$\Phi = AB \cos \theta$

where A is the area, B is the B-field and is the angle between the normal to the surface and the B-field.

3.10.1 Magnetic flux

Here's a quick definition that will give you an idea of why the B-field is often given the strange sounding name of *magnetic flux density*. Take a look at the diagram of a surface of area A in a uniform B-field (Fig. 3.10.1)

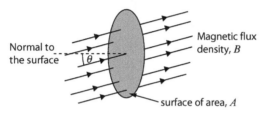

Fig. 3.10.1 Magnetic flux

In this set-up, the magnetic flux of the surface is defined as

$$\Phi = AB \cos \theta.$$

You also need to remember the unit of magnetic flux – the weber, **Wb**. If you can't remember this in an exam use a bit of common sense and write the unit as $T\ m^2$ and save yourself the possible deduction of a mark (see **Pointer**).

In many respects, the difficulty has already started because you have to imagine an area and an invisible magnetic field. However, the area will nearly always be the area of some sort of a loop of wire and won't be related to an imaginary surface.

By rearranging the equation and taking θ = zero (which it will be for nearly all A-level questions):

$$B = \frac{\Phi}{A}$$

so that the B-field is the magnetic flux divided by the area or, in other words, the *magnetic flux density* (and hence the strange name).

> **Pointer**
>
> You can often obtain an alternative unit using an equation, e.g. the unit of magnetic flux (Wb) is $m^2 \times T$ from $\Phi = AB \cos \theta$, (remember that $\cos \theta$ is a ratio and dimensionless).

quickpire

① The axis of a metal circular loop is at an angle of 27° to the B-field that passes through it and the area of the loop is $4.6 \times 10^{-2}\ m^2$. If the B-field is 0.034 T, calculate the magnetic flux of the loop.

Example

The magnetic flux through a coil of area $32\,\text{cm}^2$ is $8.7\,\mu\text{Wb}$. The B-field through the coil is uniform and always at right angles to the area enclosed by the coil. Calculate B, the magnetic flux density.

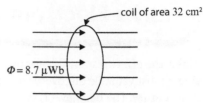

Fig. 3.10.2 Calculating flux

Answer

First, $\theta = 0$ and you can use the rearranged version of the equation shown above

$$B = \frac{\Phi}{A} = \frac{8.7 \times 10^{-6}\,\text{Wb}}{32 \times 10^{-4}\,\text{m}^2} = 27\,\text{mT}$$

3.10.2 Flux linkage

This is no more complicated than magnetic flux but refers to many loops rather than one loop.

If a coil has N loops and the magnetic flux through each loop is Φ then the total magnetic flux for all the loops is:

total magnetic flux for the whole of the coil = **flux linkage** = $N\Phi$

The unit of flux linkage is also the weber (Wb) but it can also be written as weber turn (Wb turn). The 'turn' part is not essential but is useful. Most of the time you'll be able to write

flux linkage $= N\Phi = BAN$

because $\cos\theta = 1$ and the same flux passes through each of the loops (or turns).

Example

A solenoid (cylindrical coil of wire) has 3600 turns (or loops). A uniform magnetic flux density of 3.8 mT passes through the centre of the solenoid parallel to its axis.

The solenoid has a circular cross-section of radius 5.3 cm. Calculate the flux linkage for the solenoid.

radius 5.3 cm

3.8 mT

Fig. 3.10.3 Flux linkage in a solenoid

Answer

Once again, we have $\theta = 0$ and $\cos\theta = 1$ (the 3600 loops will be much flatter than the simplified diagram), so the flux for each turn is:

$$\Phi = AB = \pi r^2 B = \pi \times 0.053^2 \times 0.0038 = 3.35 \times 10^{-5}\,\text{Wb}$$

Then you need to use the equation for the flux linkage

flux linkage $= N\Phi = 3600 \times 3.35 \times 10^{-5} = 0.12\,\text{Wb (or Wb turn)}$

≫ Pointer

There is no need to do the calculation in two parts. Just use BAN.

3.10.3 Faraday's and Lenz's laws

Faraday's law is the easier to understand (and apply). It is only a short sentence (see **Key terms**) but every single dynamo (generator, alternator or transformer) is based on this law.

Lenz's law is rather tricky – we'll look at it after a few examples using Faraday's law.

Taken together the laws can be written: $\mathcal{E}_{in} = -\dfrac{\Delta(N\Phi)}{\Delta t}$ or $-\dfrac{\Delta(BAN)}{\Delta t}$

where $\mathcal{E}_{in}$ is the induced EMF and the minus sign is the expression of Lenz's law (see **Pointer**). Note that, as you need to remember the laws of Faraday and Lenz, these equations are not included in the Data booklet.

(a) Applying Faraday's law

You should realise that there are several ways of inducing an emf from Faraday's law.

1 **Varying the B-field**
Transformers make use of this but they are not on this specification. If you study Option A you will meet inductors, which control currents using e-m induction with a varying magnetic field.

2 **Varying the area** (through some sort of motion)
Conductors moving through a magnetic field generate an emf.

3 **Varying the angle between the field and coil**
Electrical generators work using this principle (see below).

Example 1 (changing field)

The flux density of a uniform magnetic field at right angles to a circular coil of 100 turns and diameter 20 cm increases steadily from 0.05 T to 0.25 T in 50 ms. Calculate the emf, $\mathcal{E}$, induced in the coil.

Answer

$\mathcal{E} = \dfrac{\Delta(BAN)}{\Delta t} = AN\dfrac{\Delta(B)}{\Delta t}$ because A and N are constant

$= \pi \times (0.10\,m)^2 \times 100 \times \dfrac{(0.25-0.05)T}{0.05\,s}$

$= 12.6\,V$

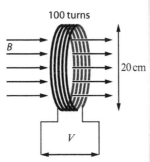
100 turns
B
20 cm
V

Fig. 3.10.4 EMF induced in a coil by a varying field

Example 2 (changing area)

A thick conductor slides along the rail tracks at $34\,\text{m s}^{-1}$ as shown. Calculate the emf induced and the current in the resistor.

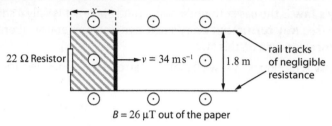

Fig. 3.10.5 Sliding conductor on rails

Answer

Start with Faraday's mathematical expression: $\varepsilon_{in} = \dfrac{\Delta(BAN)}{\Delta t}$ (and ignoring the – sign for the moment).

You should note that $N = 1$ (there's definitely only 1 loop, shaded) and that the B-field is a constant. Hence (ignoring the minus sign)

$$\varepsilon_{in} = B\frac{\Delta(A)}{\Delta t}$$

but the area A is given by $A = l \times x$, where l is the length of the conducting bar (1.8 m), so

$$\varepsilon_{in} = B\frac{\Delta(lx)}{\Delta t} = Bl\frac{\Delta(x)}{\Delta t} = Blv \text{ (see \textbf{Pointer})}$$

because $\dfrac{\Delta x}{\Delta t}$ (or $\dfrac{x}{t}$) is the speed v. Putting in the numbers gives

$$\varepsilon_{in} = Blv = 26 \times 10^{-6}\,\text{T} \times 1.8\,\text{m} \times 34\,\text{m s}^{-1} = 1.59\,\text{mV}$$

Calculating the current is left as a (fairly trivial) exercise – Quickfire 7.

(b) An alternative approach to Faraday's law

An alternative approach to the last question might also help you understand Faraday's law. You can consider the moving conductor as a cell providing the emf for the whole circuit. The free electrons within the conductor will experience a force because they are moving within a magnetic field. If you apply FLHR to these electrons they will move upward. The work done on an electron moving from the bottom of the conductor to the top is:

$$\text{Work} = \text{force} \times \text{distance} = Bqv \times l$$

(Note: The force and distance moved are in the same direction)

But thinking of the moving conductor as a cell providing the (induced) emf. The work done is also:

$$\text{Work} = \text{emf} \times \text{charge moved} = \varepsilon_{in}q$$

These two expressions for the work must give the same answer, so:

$$Bqvl = \varepsilon_{in}q \quad \text{and dividing by } q \text{ leaves} \quad \varepsilon_{in} = Blv$$

(c) Applying Lenz's law to these examples

Example 1

In Fig. 3.10.4, what is the direction of the induced current in the coil when the magnetic field is increasing?

Answer

One way of applying Lenz's law is to look at the magnetic flux linkage inside the circuit.

- The flux linkage is increasing so the induced current must oppose the increase, so
- the flux caused by the induced current must be in the opposite direction (i.e. to the left), so
- using the right hand grip rule the direction of the current must be clockwise viewed from the right (Fig. 3.10.6).

Fig. 3.10.6 Direction of induced field, B_{in} and current, I_{in}.

Example 2

In what direction is the induced current in the resistor?

Answer

Method 1 (Using FLHR):

Lenz's law says that there must be a force on the moving rod opposing the motion, i.e. to the left.

First finger (field) towards you:

Thumb (motion) to left

∴ Current in bar (second finger) must be downwards.

∴ Current in the resistor is upwards.

Method 2 (the flux linkage method):

The flux linkage in the circuit is increasing, so (Lenz's law) the flux produced by the induced current must be in the opposite direction, i.e. into the diagram. Hence the direction of the induced current must be clockwise in the loop, i.e. upwards in the resistor.

Fig. 3.10.7 In what direction is the induced current?

>> *Pointer*

Remember: Lenz's law says the induced current opposes the change that caused it.

quickꟼire

⑨ Use the same reasoning to predict the direction of the induced current if the applied magnetic field is decreasing in strength (i.e. $\Delta B < 0$).

Grade boost

In Example 2 the two ways of applying Lenz's law give the same answer – they always do!

quickꟼire

⑩ Use a third method to answer Example 2:

Consider the electrons which are moving because the bar is moving. Use FLHR to predict the direction of the force on them and hence answer the question.

(d) Fleming's right-hand rule

This is yet another way of predicting the direction of an induced current. It is only useful in the cases where a conductor is moving across a magnetic field: you cannot use it where a magnetic field is changing inside a circuit (e.g. Example 1 above).

How does it work? The same as FLHR except that you use the right hand!

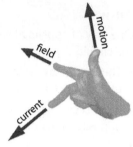

That means: <u>F</u>irst finger   <u>F</u>ield

              Se<u>C</u>ond finger   (induced) <u>C</u>urrent

              Thu<u>M</u>b   <u>M</u>otion

See Fig. 3.10.8 and compare it with Fig. 3.9.2!

Fig. 3.10.8 Fleming's right-hand rule

3.10.4 Rotating coil in a magnetic field

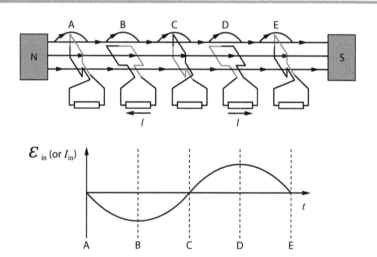

Fig. 3.10.9 Rotating generator coil

Although you don't have to do calculations based on a rotating coil dynamo or generator, you do need to be able to explain the effect of the following on the induced emf:

1 coil position   2 flux density   3 coil area   4 angular velocity

All four effects can be explained with reference to Fig. 3.10.9 and Faraday's law.

1 Coil position

This is far easier to explain in terms of flux cutting but the alternative explanation in terms of changing flux is also given.

In positions **A**, **C** and **E**, the induced emf is zero because the coil is not cutting lines of magnetic flux (the long sides of the coil are moving parallel with the field lines and so are doing no cutting). Alternatively, you can state that the flux linkage of the coil is at its maximum (because $\cos \theta = 1$). If the flux linkage is a maximum, then the rate of change of flux linkage is zero.

In positions **B** and **D** the induced emf is a maximum (and opposite) because the coil is cutting lines of magnetic flux at right angles (i.e. cutting field lines at the greatest rate). Alternatively, you can state that the flux linkage of the coil is changing at its greatest rate (this is true even though the flux linkage of the coil is zero since $\cos \theta = 0$).

2 Flux density

The induced emf is proportional to the strength of the B-field. This is easily explained because a double strength B-field results in double the lines of magnetic flux being cut. Alternatively, a doubled B-field results in a doubled magnetic flux linkage for the coil: hence a doubled rate of change.

3 Coil area

The induced emf is proportional to the coil area. This is because a doubled area results in a double magnetic flux linkage for the coil: hence a doubled rate of change. Alternatively, the larger the area of the coil the more lines of magnetic flux will be cut.

4 Angular velocity

The induced emf is proportional to the angular velocity. Again, both flux cutting and change in flux linkage approaches are valid. If the angular velocity increases, it's obvious that the rate of change of flux linkage increases in proportion, it's also obvious that the rate of cutting of flux increases.

≫ *Pointer*

Suppose you plot a graph of some variable against time. When the variable reaches its maximum or minimum value, the line plotted must be horizontal (otherwise it couldn't be a maximum or minimum because it would still be increasing or decreasing). If the line is horizontal, the gradient is zero and hence the rate of change of the variable is zero.

quickᴘıre

⑪ Explain why the induced emf is reversed in position D compared with position B.

1. (a) Calculate the magnetic flux for the flying saucer shown in the Earth's magnetic field.

    (b) The flying saucer is flying extremely quickly. Use Faraday's law to explain why the emf around the rim of the saucer is always zero if the magnetic field is uniform.

    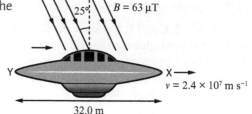

    (c) Calculate the force acting on free electrons at points X and Y (using $F = Bqv$).

    (d) Use your answers to (c) to explain why the induced emf is zero for the rim of the flying saucer.

2. (a) Calculate the flux linkage for the solenoid shown (you will need to count the number of turns).

    (b) The source of the magnetic field is switched off and the B-field drops to zero in 55 μs. Calculate the induced emf in the solenoid.

    (c) State the direction of the induced current in the solenoid (if there were a complete circuit).

3. A metal bar slides along a diverging pair of rails as shown.

    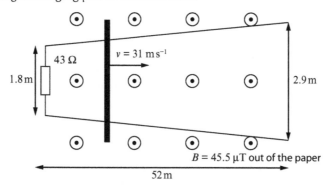

    Assuming the bar and rails have negligible resistance:

    (a) Explain why an induced current flows in the resistor.

    (b) State the direction of this induced current in the resistor and state which rule you used to obtain your answer.

    (c) Calculate:

        (i) the change in flux as the sliding conductor travels the whole 52 m from left to right

        (ii) the mean current.

    (d) Explain briefly why this is a mean current.

    (e) Sketch the following graphs for the whole 52 m of motion:

        (i) the induced current against time

        (ii) the power dissipated in the resistor against time.

4. A strong magnet drops vertically through a flat coil.

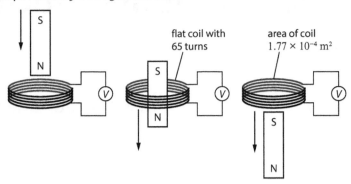

flat coil with 65 turns

area of coil 1.77×10^{-4} m²

The emf induced in the coil is recorded using a voltmeter.

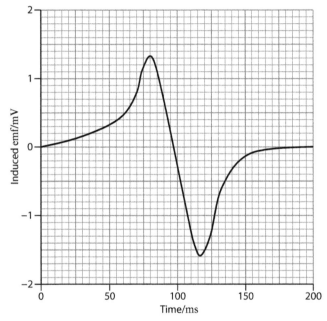

(a) Use the laws of Faraday and Lenz to explain why the measured emf varies as shown in the graph.

(b) The voltmeter is now removed and the ends of the flat coil connected so that current can flow. Sketch a graph showing the variation of force exerted by the coil on the magnet against time (no calculations are required).

(c) Use the information in the induced emf graph to estimate the maximum flux density of the falling magnet.

5. A metal wire coat hanger is held perpendicular to a uniform magnetic field of 2.8 mT. The frame is pulled apart as shown so that the area inside the coat hanger is increased from 220 cm² to 560 cm² in a time of 0.050 s. The resistance of the coat hanger loop is 0.13 Ω.

(a) Use Lenz's law to determine the direction of the current induced in the coat hanger.

(b) Calculate the mean magnitude of the current induced.

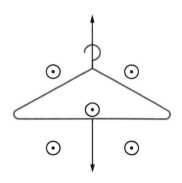

Component 3 Light and nuclei Summary

3.6 Nuclear decay

- The spontaneous nature of decay
- The properties of α, β and γ radiation; nuclear transformation equations using the $^A_Z X$ notation
- Distinguishing between α, β and γ radiation; the connections between nature, penetration and range of nuclear radiation
- Investigating radioactive decay; allowing for background radiation
- Activity, A, the decay constant, λ, and the relationship $A = \lambda N$
- The concept of half-life, $T_{\frac{1}{2}}$
- The exponential relationships
  $N = N_0 e^{-\lambda t}$ and $A = A_0 e^{-\lambda t}$
  or $N = \dfrac{N_0}{2^x}$ and $A = \dfrac{A_0}{2^x}$
  where x is the number of half-lives (not necessarily an integer)
- The use of log graphs to investigate radioactive decay; the derivation and use of $\lambda = \dfrac{\ln 2}{T_{\frac{1}{2}}}$

3.8 Nuclear energy

- The significance and use of the mass-energy equivalence, $E = mc^2$
- The application of $E = mc^2$ including the use of the equivalence $1\,u \equiv 931\,MeV$
- Calculations of nuclear binding energy and the binding energy per nucleon from data on particle masses
- The application of mass-energy conservation to particle interactions, especially fission, fusion and nuclear decay
- The binding energy per nucleon graph and its relevance for nuclear fission and fusion reactions

3.9 Magnetic fields

- Force on a wire in a B-field, $F = BIl\sin\theta$ and Fleming's left-hand rule (FLHR)
- Force on a moving charge in a B-field, $F = Bqv\sin\theta$, leading to circular motion
- Strength and shape of the B-field due to a long straight wire and a solenoid; $B = \dfrac{\mu_0 I}{2\pi a}$ and $B = \mu_0 nI$; the effect of an iron core
- Mutual forces on current-carrying conductors
- Magnetic and electric field theory applied to the deflection of particle beams and to particle accelerators (cyclotrons and synchrotrons)

3.10 Electromagnetic induction

- Magnetic flux, $\Phi = AB\cos\theta$. and flux linkage, $N\Phi$
- The laws of Faraday and Lenz and their application to calculate the magnitude and
- direction of induced emf; $\mathcal{E}_{in} = -\dfrac{\Delta(N\Phi)}{\Delta t}$
- Induced emf in conductors moving at right angles to and coils rotating in uniform B-fields
- Qualitatively relating the induced emf to the position, area and angular speed of a rotating coil and the magnetic flux density

Specified practical work

- Investigation of the force on a current in a magnetic field
- Investigation of magnetic flux density using a Hall probe

Component 3 Options[1]

Knowledge and Understanding

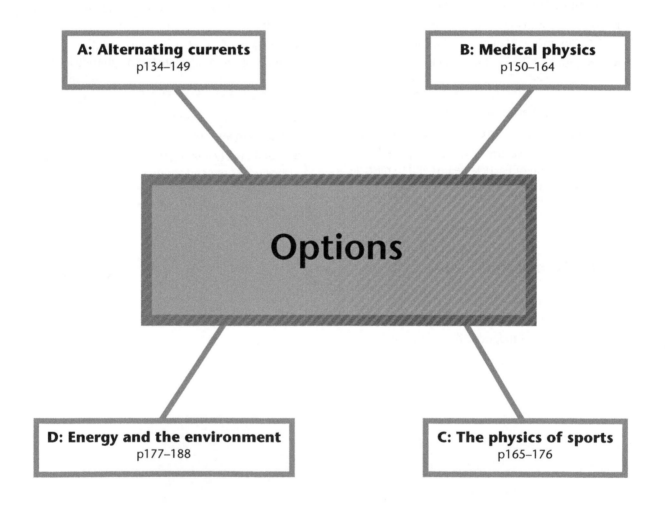

A: Alternating currents
p134–149

B: Medical physics
p150–164

Options

D: Energy and the environment
p177–188

C: The physics of sports
p165–176

[1] You should study **one** option

Revised it!

Basic notes · Good grasp · Fully revised

A: Alternating currents

emf generation by rotating coils in magnetic fields; AC currents and pds; rms values and their use in power dissipation; measurements using an oscilloscope; I/V relationships for resistors, capacitors and inductors; reactance and impedance; RCL circuits, current, pd and power; phasor analysis; resonance and the Q factor.

p134–149

B: Medical physics

X rays: their nature, production and use in diagnosis and therapy; imaging techniques; attenuation.
Ultrasound: generation and detection; A and B scans; acoustic impedance and coupling medium; Doppler scans
Magnetic resonance imaging; principles and use in diagnosis.
Nuclear medicine: dose, equivalent dose and effective dose; radiotracers; PET scanning; the gamma camera.

p150–164

C: The physics of sports

Rotational dynamics applied to sports: moment of inertia, torque, angular acceleration, angular momentum and rotational kinetic energy.
Newton's laws applied to sports; collisions; coefficient of restitution; conservation of momentum, energy and angular momentum.
Projectile motion applied to sports.
Bernoulli's equation; the drag equation and the drag coefficient.

p165–176

D: Energy and the environment

The effect of atmopsheric gases on the temperature of the Earth; the greenhouse effect including anthropogenic effects; Archimedes' principle and the effect of melting ice on sea levels.
Energy sources: solar radiation; photovoltaic cells; wind, tidal hydroelectric and pumped storage power; nuclear fission and fusion principles; fuel cells.
Thermal conduction; the coefficient of thermal conductivity; U values.

p177–188

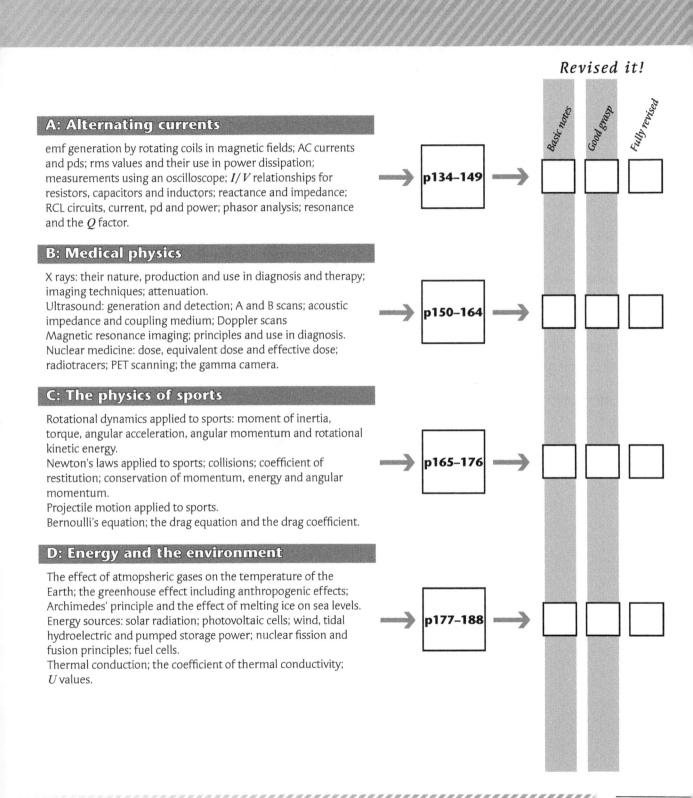

Option A: Alternating currents

A.1 A rotating coil in a uniform magnetic field

This forms the basis of all generators or dynamos. They are responsible for the generation of electricity in all power stations but also ensure that batteries in all motorised road vehicles are 'charged' continuously as they move (or brake for Kinetic Energy Recovery Systems, KERS).

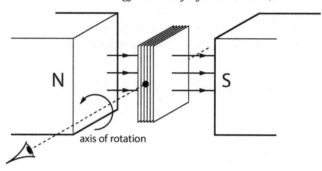

axis of rotation

Fig. A1 Coil in a uniform magnetic field

As the description and Fig. A1 suggests, an ideal generator consists of a flat coil rotating in a uniform B-field. A simplified 2D image is shown below (Fig. A2).

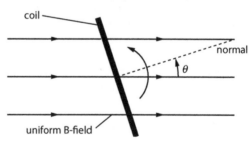

Fig. A2 Flux linkage in a rotating coil

From the definition of flux linkage it should be reasonably clear that the instantaneous flux linkage of the coil is $BAN \cos \theta$, where A is the cross-sectional area of the coil and N is the number of turns. If the coil rotates at a constant angular velocity (ω) and if the angle θ starts at 0 when the time, $t = 0$, then the angle will be given by

$$\theta = \omega t$$

Hence, the flux linkage can be written as $BAN \cos \omega t$

Using Faraday's law, the induced emf is the rate of change of flux linkage and can then be calculated:

$$\mathcal{E}_{in} = -\frac{d}{dt}(N\Phi) = -\frac{d}{dt}(BAN\cos\omega t) = -(\omega BAN(-\sin\omega t)) = \omega BAN\sin\omega t$$

Don't worry if you're not doing A-level Maths and didn't quite follow that last bit because you don't have to be able to derive the equation, only use it. The good news is that both the equation for the flux linkage and the equation for the induced emf are in the Data booklet.

If you remember that the maximum value of sin (and cos alike) is 1, then you can write:

$$\mathcal{E}_{peak} = \omega BAN$$

Example

A high-power generator supplies $1240\,MW$ of peak electrical power at a peak output pd of $30\,kV$. It has a uniform magnetic flux density of $0.11\,T$ and its rotating coil rotates at a rate of 3000 rpm and has a cross-sectional area of $26\,m^2$. Calculate:

(i)   the number of turns in the rotating coil

(ii) the peak output current.

Answer

(i)   First we must convert 3000 revolutions per minute to Hz:

$$\text{frequency} = \frac{3000\text{ revolutions}}{60\,s} = 50\text{ Hz}$$

and we need to remember the relationship between angular velocity and frequency:

$$\omega = 2\pi f$$

giving

$$\mathcal{E}_{peak} = \omega BAN = 2\pi f BAN = 30\,000\text{ V}$$

rearranging and substituting gives

$$N = \frac{\mathcal{E}_{peak}}{2\pi f BA} = \frac{30\,000}{2\pi \times 50 \times 0.11 \times 26} = 33\text{ turn}$$

(ii) Calculating the peak current is easier:

$$P = IV \quad\rightarrow\quad I = \frac{P}{V} = \frac{1240\,MV}{30\,kA} = 41\text{ kA}$$

quicKfire

① Obtain an expression for the induced emf in a rotating coil when the time, $t = \frac{T}{4}$ (a quarter of a period).

quicKfire

② State or calculate the flux linkage of a rotating coil when $t = \frac{T}{4}$.

≫ **Pointer**

Note how the flux linkage and induced emf are 90° ($\pi/2$ rad) out of phase. Hence, the flux is zero when the emf is a maximum and vice versa. This is very similar to the relationship between position and velocity in shm.

≫ **Pointer**

mega divided by kilo is $\frac{10^6}{10^3}$ i.e. $10^3 = $ kilo.

A.2 Alternating current and the root mean square (rms)

All circuit analysis so far has been based on DC circuits but it's also important to have some idea of the very basics of AC circuits.

All alternating pds will be sinusoidal in this section. Here's a graph of what the sinusoidally varying pd entering your home from the national grid looks like.

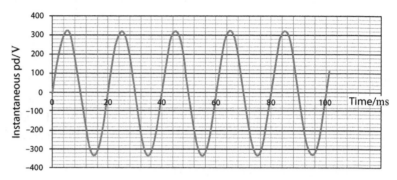

Fig. A3 Mains voltage

You should see that the period of the voltage is $20\,\text{ms}$ and that this corresponds to a frequency of $50\,\text{Hz}$ (remember $f = \dfrac{1}{T}$). However, one unfamiliar thing that you should notice is that the peak pd is around $325\,\text{V}$ and that this is different from the standard $230\,\text{V}$ that you associate with the electricity supply of your house. The reason why $230\,\text{V}$ is associated with the above graph is not at all obvious at first but it's related to the power dissipation.

Consider the following simple circuit.

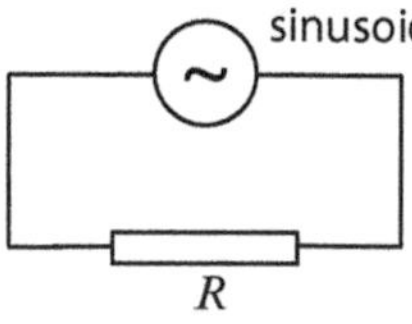

The instantaneous power dissipated in the resistor will be given by

$$P = \frac{V^2}{R}$$

Fig. A4 Resistive AC circuit

where V is the instantaneous pd. Because the pd is varying quickly we're not concerned with instantaneous power but rather with the mean power. That means you have to obtain the mean value of $\dfrac{V^2}{R}$ which means obtaining the mean value of V^2. Very similar to the concept of the root mean square velocity in Section 1.7, we use a root mean square pd value which is written as V_{rms}. The mean power dissipated in the resistor is given by

$$\langle P \rangle = \frac{V_{\text{rms}}^{\;2}}{R}$$

> ## » Pointer
> The root mean square pd V_{rms} of a sinusoidally varying pd ($V_0 \sin \omega t$) is the mean value of $V_0^2 \sin^2 \omega t$. The mean value of $\sin^2 \omega t = 0.5$ and this makes sense when you think about it. So,
> $$V_{\text{rms}}^{\;2} = V_0^2 \times \frac{1}{2}$$
> and hence,
> $$V_{\text{rms}} = \frac{V_0}{\sqrt{2}}.$$

> ## » Pointer
> $1/\sqrt{2} = 0.707$ so the rms pd is always 70.7% of the peak pd.

> ## » Pointer
> The root mean square induced emf for a rotating coil will be given by
> $$V_{\text{rms}} = \frac{1}{\sqrt{2}} \omega B A N$$

Due to the sinusoidal variation of the pd, the rms pd (V_{max}) is given by

$$V_{rms} = \frac{V_0}{\sqrt{2}} \text{ (see Section M.4.1)}$$

Example

The rms pd supplied to a house is 230 V. Calculate the peak pd (V_0).

Answer

$$V_{rms} = \frac{V_0}{\sqrt{2}} \quad \rightarrow \quad V_0 = \sqrt{2}V_{rms} = 1.4142 \times 230 = 325\,\text{V}$$

which explains the unfamiliar peak pd of 325 V shown in the earlier graph. The relationship is also true for the root mean square current I_{rms}

$$I_{rms} = \frac{I_0}{\sqrt{2}}$$

and all the expressions for electrical power are true when the rms values are used

$$\langle P \rangle = I_{rms}V_{rms} = \frac{V_{rms}^2}{R} = I_{rms}^2 R$$

Example

Hair straighteners are rated 26 W and operate from a peak pd of 340 V. Calculate:

(i) the rms pd

(ii) the rms current

(iii) the peak current

(iv) the resistance of the straighteners.

Answer

(i) $V_{rms} = \dfrac{V_0}{\sqrt{2}} = \dfrac{340}{1.4142} = 240\,\text{V}$

(ii) using $\langle P \rangle = I_{rms}V_{rms} \quad \rightarrow \quad I_{rms} = \dfrac{\langle P \rangle}{V_{rms}} = \dfrac{26}{240} = 0.11\,\text{A}$

(iii) $I_{rms} = \dfrac{I_0}{\sqrt{2}} \quad \rightarrow \quad I_0 = \sqrt{2}I_{rms} = 1.4142 \times 0.11 = 0.15\,\text{A}$

(iv) using $R = \dfrac{V}{I}$ you can use either the peak values or the rms values of the pd and current

$$R = \frac{V}{I} = \frac{240}{0.11} = 2200\,\Omega$$

quickfire

③ The peak current supplied to a toaster is 5.6 A. Calculate the root mean square current.

quickfire

④ The rms pd supplied to an electrical device is 23.0 V. Calculate the peak pd.

quickfire

⑤ A kettle element has a resistance of 19.6 Ω and the peak current supplied to it is 16.1 A. Calculate:
i) the mean power supplied to the kettle
ii) the rms pd supplied to the kettle.

quickfire

⑥ An electric heater has a rms current of 8.3 A and a peak pd across it of 340 V. Calculate its mean output power.

A.3 The oscilloscope

Gone are the days that A-level students needed to know how an oscilloscope worked and this makes sense now that freeware software can turn your PC and soundcard into an oscilloscope for free. However, the specification does state clearly that you should know how to use an oscilloscope.

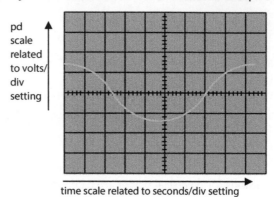

Fig. A5 Oscilloscope screen

Take a look at the oscilloscope trace showing a sinusoidally varying pd. You should be able to tell that the 'amplitude' of the trace is about 1.4 squares (vertically) and that the 'wavelength' of the trace is about 9.4 squares (horizontally).

Now to understand the full significance of the trace you need to use the volts / div and seconds / div settings.

Looking at the VOLTS / DIV setting (also called the Y-gain) in Fig. A6 you should note that the setting is 50 mV per division. This means that the height of each square on the oscilloscope represents 50 mV. Hence, the peak pd of the trace on the screen is

$$1.4 \times 50 = 70\,\text{mV}$$

Looking at the SEC / DIV setting (usually called the time base), the knob is set to .5 ms. This means that each 1 cm division on the oscilloscope represents 0.5 ms. Hence, the period of the waveform shown on the oscilloscope is

$$9.4 \times 0.5 = 4.7\,\text{ms}$$

Hence the frequency f is given by: $f = \dfrac{1}{T} = \dfrac{1}{0.0047\,\text{s}} = 210\,\text{Hz}$

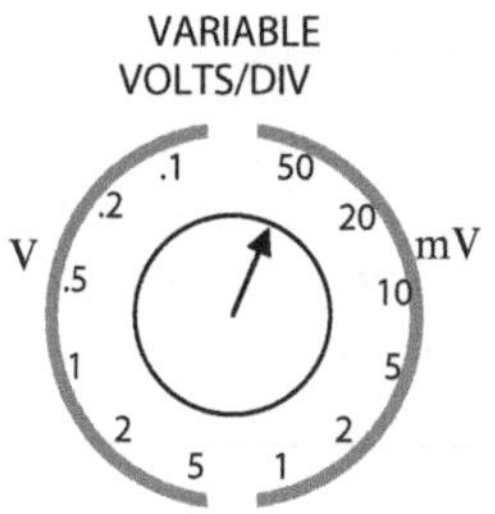

≫ Pointer

The oscilloscope trace is essentially a voltage–time graph.

≫ Pointer

On an oscilloscope screen the 1 cm squares are called divisions. Note that the *small* divisions of the squares on the oscilloscope screen are 0.2 squares.

Fig. A6 Oscilloscope Y-gain

Fig. A7 Oscilloscope time base

Example: a nasty question that an examiner could ask is:

Draw the trace on the oscilloscope screen for the same input waveform when the Y-gain is adjusted to 20 mV/div and the time-base to 1 ms/div.

Answer

To find the height of the trace,
peak pd = 70 mV, hence

$$\text{max height} = \frac{70\,\text{mV}}{20\,\text{mV}} = 3.5 \text{ squares}$$

Also, period = 4.7 ms, hence

$$\text{wave width} = \frac{4.7\,\text{ms}}{1\,\text{ms}} = 4.7 \text{ squares}$$

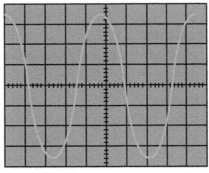

Fig. A8 AC oscilloscope trace

(a) Using an oscilloscope

This is easier than you might think. Once you've input your pd into the oscilloscope you simply twiddle the VOLTS / DIV until your signal is the maximum height that fits on the screen. Then you twiddle the SEC / DIV until you have a few complete waveforms on the screen.

Measuring DC voltage

Again, this is easier than you might think. A constant pd just gives a horizontal line on the oscilloscope screen. In the following example, the oscilloscope has been adjusted so that the zero volt level is not in the centre of the oscilloscope screen.

Example

Calculate the two DC voltages represented by the top and bottom lines on the oscilloscope screen.

The Y-gain setting of the oscilloscope is 5 mV/div.

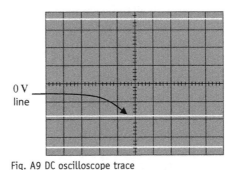

0 V line

Fig. A9 DC oscilloscope trace

Answer

The top line is 5.3 squares above the zero-volt line, hence

$$\text{pd} = 5.3 \times 5 = 26.5\,\text{mV}$$

The bottom line is 1.8 squares below the zero-volt line, hence

$$\text{pd} = -1.8 \times 5 = -9.0\,\text{mV}$$

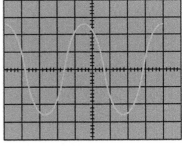
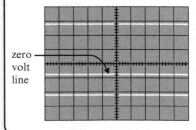

(b) Measuring currents using an oscilloscope

This is not directly possible because an oscilloscope measures pd. However, the current can be calculated if you know the device for which you're measuring the pd.

Example,

In the last DC voltage example the pds were measured across a 680 Ω resistor. Calculate the current in the resistor corresponding to the pds shown.

Answers

$$I = \frac{V}{R} = \frac{0.0265}{680} = 39 \ \mu A$$

and

$$I = \frac{V}{R} = \frac{0.009}{680} = 13 \ \mu A$$

Key Term

Phasor = a rotating vector whose length represents the rms (or peak) value of a pd and whose direction represents phase. It can also be used to represent values of resistance, reactance and impedance.

>> **Pointer**

For A-level Physics, you don't need to know the reason that phasor diagrams work – only how to apply the analysis.

>> **Pointer**

(for mathematicians)
The current in a LCR circuit is $I = I_0\sin\omega t$ but the pd **applied** to an inductor is $V_L = L\dfrac{dI}{dt}$. Hence,

$$V_L = L\frac{d}{dt}(I_0\sin\omega t)$$

$$= \omega L I_0\cos\omega t$$

$$= \omega L \times I_0\sin(\omega t + \frac{\pi}{2})$$

So, for an inductor, the pd is $\dfrac{\pi}{2}$ ahead of the current.

A.4 Phasor diagrams and AC circuits

You need to know how to apply AC circuit analysis to series circuits involving capacitors, inductors and resistors. This is done using vector or phasor diagrams.

Consider the *LCR* circuit (Fig. A10) – it's called an *LCR* circuit because it has an inductor (L), capacitor (C) and resistor (R).

Similar to a DC circuit, the current is the same through all three components but the pds are different. Added to this, however, is the complexity of phase differences. For an inductor, the pd across it is 90° in front of the current (see **Pointer**).

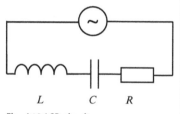

Fig. A10 LCR circuit

For a capacitor, the pd is 90° behind the current (see **Pointer**). However, a resistor is simpler, the current and pd are always in phase for a resistor. The phasor is drawn as follows:

First draw the resistor pd as a horizontal vector, you don't need to draw the current vector but it's been added for completeness. Then draw V_L as going vertically upward. This signifies that V_L is 90° ahead of V_R (increasing phase goes anticlockwise). Draw V_C going vertically downwards, signifying that V_C is 90° behind V_R.

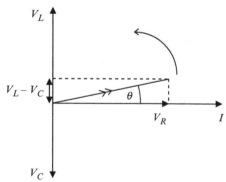

Fig. A11 LCR phasor diagram

You obtain the resultant in the same way as normal vectors – first obtain $V_L - V_C$ then obtain the resultant of $V_L - V_C$ and V_R. Hence, the resultant is the phasor shown in the diagram with the double arrow. The magnitude of the resultant is V_S and the phase angle θ can be obtained from

$$V_S = \sqrt{(V_L - V_C)^2 + V_R^2)} \quad \text{and} \quad \tan\theta = \frac{V_L - V_C}{V_R}$$

The resultant (V_S) shown in the phasor diagram is the supply pd (hence, V_S). In general there's a phase difference between the applied pd and the current: it is represented by θ in the above diagram.

In order to calculate the current and the individual pds, you need to know the 'effective resistance' of the inductor and capacitor. This 'effective resistance' of the inductor is called the **reactance** (see **Key terms**) of the inductor and is given the symbol X_L. Likewise, the capacitor also has a reactance that's given the symbol X_C. Unlike resistance, these reactances are frequency dependent and are given by the following formulae that appear in the Data booklet (also see **Pointers**).

$$X_L = \omega L \quad \text{and} \quad X_C = \frac{1}{\omega C}$$

By definition (see **Key terms**), the pd's across the inductor and capacitor can be written:

$$V_L = IX_L = I\omega L \quad \text{and} \quad V_C = IX_C = \frac{I}{\omega C}$$

When combined with $V_R = IR$, you can obtain an expression for the 'effective resistance' of the LCR combination.

From above,

$$V_S = \sqrt{(V_L - V_C)^2 + V_R^2} = \sqrt{\left(I\omega L - \frac{I}{\omega C}\right)^2 + (IR)^2}$$

$$= I\sqrt{\left(\omega L - \frac{1}{\omega C}\right)^2 + R^2}$$

leading to

$$Z = \frac{V_S}{I} = \sqrt{\left(\omega L - \frac{1}{\omega C}\right)^2 + R^2}$$

This is where yet another new term is introduced. The name given to the 'effective resistance' of the *LCR* combination is the **impedance**, Z (see Key term). By definition, $\frac{V_S}{I}$ is the impedance of the *LCR* combination (V_S and I both represent rms values).

Hence the impedance of an LCR circuit is: $Z = \sqrt{\left(\omega L - \frac{1}{\omega C}\right)^2 + R^2}$

》 Pointer

(for mathematicians)
For a capacitor in an LCR circuit,

$$\frac{dQ}{dt} = I = I_0 \sin\omega t$$

Integrating this:

$$Q = \int I_0 \sin\omega t \, dt = \frac{1}{\omega} I_0 \cos\omega t$$

But $Q = CV_C$, so

$$V_C = \frac{1}{\omega C} I_0 \cos\omega t$$

$$= \frac{I_0}{\omega C} \sin\left(\omega t - \frac{\pi}{2}\right)$$

Hence, for a capacitor, the pd is $\frac{\pi}{2}$ behind the current.

Key Terms

The **reactance** of an inductor,

$X_L = \frac{V_{rms}}{I_{rms}}$, where V_{rms} and I_{rms} are the rms values of the pd across and the current in the inductor. It is equal to ωL (or $2\pi f L$).

UNIT: Ω

The **reactance** of a capacitor,

$X_C = \frac{V_{rms}}{I_{rms}}$, where V_{rms} and I_{rms} are the rms values of the pd across and the current in the capacitor. It is equal to $\frac{1}{\omega C}$ (or $\frac{1}{2\pi f C}$).

UNIT: Ω

Example

Calculate:

(i)   the reactance of the inductor

(ii) the reactance of the capacitor

(iii) the impedance of the circuit

(iv) the rms current

(v) the phase difference between the supply pd and the current

(vi) the frequency at which the reactances of the inductor and capacitor are equal.

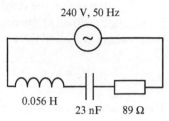

Fig. A12 Series AC circuit example

240 V, 50 Hz — *0.056 H* — *23 nF* — *89 Ω*

Answer

(i) $X_L = \omega L = 2\pi f L = 2\pi \times 50 \times 0.056 = 17.6\,\Omega$

(ii) $X_C = \dfrac{1}{\omega C} = \dfrac{1}{2\pi f C} = \dfrac{1}{2\pi \times 50 \times 23 \times 10^{-9}} = 138\,k\Omega$

(iii) $Z = \sqrt{\left(\omega L - \dfrac{1}{\omega C}\right)^2 + R^2} = \sqrt{(17.6 - 138 \times 10^3)^2 + 89^2} = 138\,k\Omega$

(iv) $I = \dfrac{V_{rms}}{Z} = \dfrac{240}{138 \times 10^3} = 1.7\,mA$

(v) this is tricky but the answer is very close to $-90°$ (or) because the circuit is, more or less, behaving like a capacitor (see above figures where the capacitor reactance is dominant, i.e. $Z \approx X_C$).

Mathematically:

$$\tan\theta = \dfrac{V_L - V_C}{V_R} = \dfrac{X_L - X_C}{R} = \dfrac{17.6 - 138\,000}{89} = -1550$$

$\therefore\quad \theta = \tan^{-1}(-1550) = 89.96°$

(vi) The reactances are equal when $\omega L = \dfrac{1}{\omega C}$, $\therefore \omega = \dfrac{1}{\sqrt{LC}}$

$$f = \dfrac{1}{2\pi\sqrt{LC}} = \dfrac{1}{2\pi\sqrt{0.056 \times 23 \times 10^{-9}}}\,Hz = 4.4\,kHz$$

Again, notice in the last example that the reactance of the capacitor ($138\,k\Omega$) is far greater than the reactance of the inductor ($17.6\,\Omega$) at 50 Hz and also far greater than the resistance 89 kΩ. This meant that the final impedance was almost exactly the reactance of the capacitor. However, at a higher frequency of 4.4 kHz the reactances of the capacitor and inductor were equal.

This is because the reactance of the capacitor $\left(\dfrac{1}{\omega C}\right)$ decreases with frequency whereas the reactance of the inductor, ωL increases with frequency.

Pointer

Remember the relationship between ω and f i.e. $\omega = 2\pi f$ and $f = \dfrac{\omega}{2\pi}$.

quickfire

⑩ Draw a phasor diagram for an *LR* circuit (it's the phasor diagram for an *LCR* circuit but without the capacitor phasor).

quickfire

⑪ Draw a phasor diagram for a *CR* circuit.

Pointer

$\dfrac{V_L - V_C}{V_R} = \dfrac{IX_L - IX_C}{IR}$

$= \dfrac{X_L - X_C}{R}$

Fig. A13 shows how the resistance, reactance and impedance of the circuit vary with frequency.

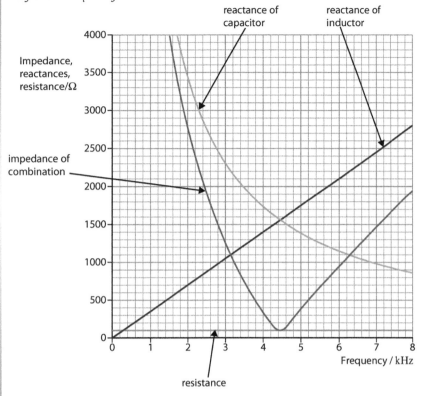

Fig. A13 Variation of R, X and Z with f

Notice how the impedance of the circuit becomes a minimum at 4.4 kHz. This brings us nicely to the next topic – resonance.

A.5 Resonance in an *LCR* circuit

Looking carefully at the last graph there are three crucial things to notice and remember:

1. The impedance is a minimum at resonance (4.4 kHz for the circuit in Fig. A13).
2. When the impedance is a minimum, the lines for the reactances of the capacitor and the inductor cross, i.e. the reactances are equal.
3. The minimum value of the impedance is the resistance.

So, for resonance, the important condition is that the reactances of the inductor and capacitor are equal:

$$\text{i.e. } X_L = X_C \quad \text{or} \quad \omega L = \frac{1}{\omega C}.$$

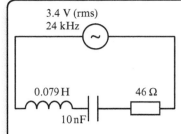

⑫ Calculate:
  i)   the reactance of the inductor
  ii) the reactance of the capacitor
  iii) the impedance of the circuit
  iv) the rms current
  v)   the phase difference between the V_S and the current
  vi) the frequency when the reactances of the inductor and capacitor are equal.

>> *Pointer*

Actually, it is easier to derive

$$Z = \sqrt{\left(\omega L - \frac{1}{\omega C}\right)^2 + R^2}$$

from the phasor diagram of resistance and reactance.

 Grade boost

The three essential points of resonance are important. Understand them and you'll do well. Resonance – minimum impedance, $X_L = X_C$ and $Z = R$.

Grade boost

Know your resonance LCR circuit, it comes up very regularly.

» Pointer

$f_0 = \dfrac{\omega_0}{2\pi}$ is the usual symbol for the resonance frequency.

but this is also reasonably obvious from the equation for the impedance.

$$Z = \sqrt{\left(\omega L - \frac{1}{\omega C}\right)^2 + R^2}$$

You want the impedance to be as small as possible in order to obtain a large current. Because the resistance R is a constant, the only way to achieve this minimum impedance is to have the two terms in the bracket cancel each other out. This gives:

$$Z = \sqrt{\left(\omega L - \frac{1}{\omega C}\right)^2 + R^2} = \sqrt{(0)^2 + R^2} = \sqrt{R^2} = R$$

and this explains why the minimum impedance in the graph is actually the resistance of the resistor R. The expression for the resonance frequency has already been derived in the last example of an LCR circuit but here it is again to make sure you learn it:

$$\omega_0 L = \frac{1}{\omega_0 C} \quad \rightarrow \quad \omega_0^2 = \frac{1}{LC} \quad \rightarrow \quad \omega_0 = \frac{1}{\sqrt{LC}} \quad \rightarrow \quad f_0 = \frac{1}{2\pi\sqrt{LC}}$$

Example

The circuit is at resonance.

Calculate:

(i)   the current

(ii) the inductance of the inductor

(iii) the pd across the capacitor and inductor

(iv) the phase difference between the applied pd and the current.

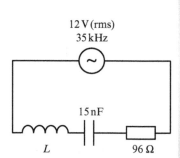

12 V (rms)
35 kHz

15 nF

L 96 Ω

Answer

(i) Because the circuit is at resonance the reactances of the inductor and capacitor cancel and $Z = R$. Hence,

$$I = \frac{V}{Z} = \frac{V}{R} = \frac{12}{96} = 0.125\,\text{A}$$

(ii) This is trickier and you need to use the resonance condition

$$X_L = X_C \qquad \omega L = \frac{1}{\omega C} \qquad L = \frac{1}{\omega^2 C}$$

$$L = \frac{1}{(2\pi f)^2\, C} = \frac{1}{(2\pi \times 35 \times 10^3)^2 \times 15 \times 10^{-9}} = 1.38\,\text{mH}$$

(iii) $V_L = IX_L = I\omega L = 0.125 \times 2\pi \times (35 \times 10^3) \times 0.00138 = 38\,V$

You don't need to calculate the pd across the capacitor because it must be equal to 38 V. You should check this using $V_C = IX_C$.

$$V_C = IX_C = \frac{I}{\omega C} = \frac{I}{2\pi f C}$$

$$= \frac{0.125}{2\pi \times 35\,000 \times 15 \times 10^{-9}} = 38\,V$$

(iv) The resultant phasor at resonance is the resistance. This means that the current and applied pd are in phase and the phase angle is zero.

An astonishing result in that last example is that the pd across the inductor and capacitor is 38 V, which is larger than the applied pd of 12 V. Although this is surprising, you must remember that the phase difference between the pd across the inductor and the pd across the capacitor is 180° – they are in anti-phase. When the pd across the inductor is positive, the pd across the capacitor is equal but negative, leaving the whole of the applied pd across the resistor.

A.6 The Quality (Q) factor of a resonance circuit

The quality (Q) factor of an LCR circuit is related to the sharpness of the resonance curve. A high Q factor gives a sharp resonance curve while a low Q factor gives a broad resonance curve (see the diagram below with $Q = 8$, $Q = 2$ and $Q = 0.5$).

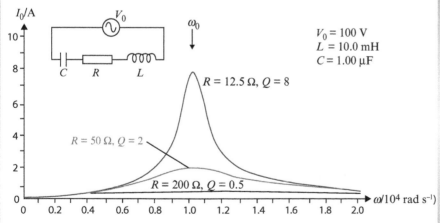

Fig. A.14 Resonance curves

quickfire

⑬ A 43 mH inductor and an 18 Ω resistor are connected in series with a capacitor across a 2.4 V, 12 kHz power supply. The circuit is at resonance. Calculate:

i)   the current

ii) the capacitance of the capacitor

iii) the pd across the capacitor and inductor

iv) the phase difference between the power supply pd and the current.

quickfire

⑭ The frequency of the supply is changed in Quickfire 13, first to 6 kHz and then to 24 kHz. Calculate the current at both these new frequencies and explain briefly why they are equal.

Key Term

Quality (Q) factor of an LCR circuit $\dfrac{\omega_0 L}{R} = \dfrac{1}{\omega_0 CR}$. It is a measure of the sharpness of the resonance curve – the larger the Q factor the sharper the resonance curve.

quickfire

⑮ Calculate the maximum and minimum resonance frequencies of the circuit (use the minimum and maximum values of the variable capacitor).

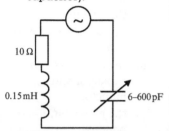

quickfire

⑯ Calculate the maximum and minimum Q factors for the circuit in Quickfire 15.

quickfire

⑰ Write down the maximum and minimum resonance frequencies of the circuit.

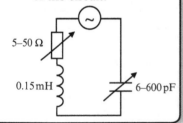

quickfire

⑱ Calculate the maximum and minimum Q factors for the circuit in Quickfire 17.

The **main** factor determining the Q factor of the circuit is the resistance of the circuit because it is the resistance that dissipates energy away from the circuit. This is similar to pushing a swing back and forth – if there's a lot of friction taking energy away from the swing it's difficult to achieve a high amplitude and 'sharp' resonance. The easiest way to define the Q factor is:

$$Q = \frac{\text{rms pd across inductor at resonance}}{\text{rms pd across resistor at resonance}}$$

As the capacitor and inductor have equal reactance at resonance, the Q factor can also be written:

$$Q = \frac{\text{rms pd across capacitor at resonance}}{\text{rms pd across resistor at resonance}}$$

These definitions lead to: $Q = \dfrac{I\omega_0 L}{IR}$, hence $Q = \dfrac{\omega_0 L}{R}$.

You should also be able to derive: $Q = \dfrac{1}{\omega_0 RC}$ and $Q = \dfrac{1}{R}\sqrt{\dfrac{L}{C}}$

So you have three expressions for the Q factor (only the first appears in the Data booklet).

Now consider this circuit:

These values for R, C, L give you a quick exercise in dealing with powers of 10.

You should obtain the following figures:

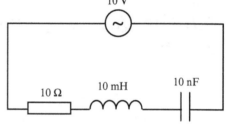

Fig. A15 An easy Q-factor circuit

$$\omega_0 = \frac{1}{\sqrt{LC}} = \frac{1}{\sqrt{10^{-2} \times 10^{-8}}} = \frac{1}{\sqrt{10^{-10}}} = 10^5 \text{ s}^{-1}$$

and

$$Q = \frac{\omega_0 L}{R} = \frac{10^5 \times 10^{-2}}{10} = 100$$

You should also be able to show (in your head!) that the current flowing at resonance is 1.0 A. (Hint: Consider the reactances of the inductor and the capacitor at resonance.)

All seems nice and straight forward until you look at the pd across the capacitor or inductor.

$$V_L = I\omega_0 L = 1 \times 10^5 \times 10^{-2} = 1000 \text{ V}$$

How can you have $1000\,V$ across the inductor (and capacitor) when the supply voltage is only $10\,V$? There's no simple answer to this question but a better understanding can be drawn from considering another type of resonance. Again, consider a swing with very little friction. You only need to provide a small push regularly in order to obtain a large amplitude – you might only be pushing the swing for a distance of $30\,cm$ each swing but the amplitude of oscillation could easily be $2\,m$.

Example

For the circuit in Fig. A16:

(i) Calculate the resonance frequency.

(ii) Calculate the Q factor.

(iii) State what happens to the resonance curve if the resistance doubles.

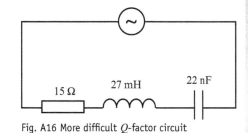

Fig. A16 More difficult Q-factor circuit

(iv) State what happens to the resonance curve when the inductance is doubled.

(v) State what happens to the resonance curve when the capacitance is doubled.

Answer

(i) $f = \dfrac{1}{2\pi\sqrt{LC}} = \dfrac{1}{2\pi\sqrt{0.027 \times 22 \times 10^{-9}}} = 6530\ \text{Hz}$

(ii) $Q = \dfrac{\omega_0 L}{R} = \dfrac{2 \times 6530 \times 0.027}{15} = 17.$

(iii) $Q = \dfrac{\omega_0 L}{R}$. As ω_0 is independent of R, Q halves and the resonance curve becomes wider.

(iv) $Q = \dfrac{\omega_0 L}{R} = \dfrac{1}{R}\sqrt{\dfrac{L}{C}}$: so Q is multiplied by $\sqrt{2}$ and the frequency of the peak of the resonance curve is divided by $\sqrt{2}$.

(v) $Q = \dfrac{\omega_0 L}{R} = \dfrac{1}{R}\sqrt{\dfrac{L}{C}}$: so both Q and the resonance frequency are divided by $\sqrt{2}$ and the resonance curve becomes wider.

1. A rectangular rotating coil rotates in a uniform magnetic field of 0.24 T. Its dimensions are 5.0 cm × 4.0 cm and a graph showing the time varying flux linkage of the coil is shown.

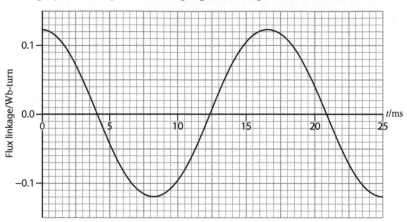

(a) Calculate the number of turns in the rectangular rotating coil.

(b) Plot a graph of the induced emf in the coil.

(c) Calculate the ratio $\dfrac{\text{peak power}}{\text{rms power}}$ if the rotating coil were to provide a current.

2 A sinusoidal pd is supplied to a kettle. The **peak** pd is 330 V and the **rms** current is 11.2 A. Calculate:

(a) the rms pd

(b) the peak current

(c) the rms power

(d) the peak power

(e) the minimum instantaneous power.

3. (a) The oscilloscope trace shown shows the output emf from a rotating coil generator. The oscilloscope settings are 5 V/div and 20 μs/div. Calculate:

(i) the peak induced emf

(ii) the rms induced emf

(iii) the frequency of rotation of the coil.

(b) The coil has dimensions 4.8 cm × 2.2 cm and has 20 turns. Use your answer to (a)(i) to calculate the magnetic flux density in which the coil rotates.

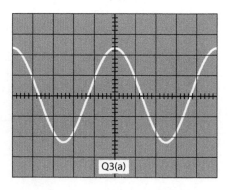

Q3(a)

(c) A second student analyses the same induced emf and obtains the following trace. Calculate:

   (i) the Y-gain setting of the oscilloscope

   (ii) the time-base setting of the oscilloscope.

(d) Copy the diagram and add a line on the screen that corresponds to a pd of $0\,V$.

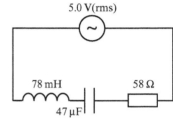

Q3(c)

4 (a) For the circuit shown, calculate:

   (i) the resonant frequency (f_0)

   (ii) the rms current at resonance

   (iii) the quality (Q) factor

   (iv) the rms pd across the capacitor at resonance

   (v) check your answer to (iv) by another method.

(b) The frequency of the supply is increased to $1.2\,f_0$. Calculate:

   (i) the impedance of the circuit

   (ii) the rms current in the circuit

   (iii) the rms pd across the resistor

   (iv) the rms pd across the inductor

   (v) the rms pd across the capacitor

   (vi) the phase angle between the supply pd and the current.

5.0 V(rms)

78 mH   58 Ω

47 µF

(c) Without further calculation, state how your answers to (b) would differ (or not) if the frequency were decreased to $\dfrac{f_0}{1.2}$.

5 (a) For the circuit shown, calculate:

   (i) the maximum and minimum current at resonance

   (ii) the maximum and minimum resonance frequency

   (iii) the maximum and minimum Q factor.

(b) Both the resistance and capacitance are set to the minimum values ($8\ \Omega$ and $3\ \mu F$) and the frequency of the $15.0\,V$ supply is set to $6.5\,kHz$. Calculate:

   (i) the impedance of the circuit

   (ii) the rms current in the circuit

   (iii) the rms pd across the resistor

   (iv) the rms pd across the inductor

   (v) the rms pd across the capacitor

   (vi) the phase angle between the supply pd and the current.

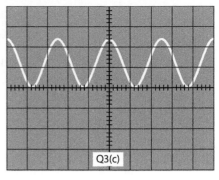

15.0 V(rms)

8–40 Ω

0.35 mH   3–30 µF

(c) The power supply is earthed and so are the inputs to an oscilloscope. Explain very briefly why the pd across the capacitor or the resistor can be displayed on the oscilloscope but the circuit will not behave correctly if the pd across the inductor is attempted.

Option B: Medical physics

This option is split into the following areas:

- X-rays – production, properties and uses
- Ultrasound – production, properties and uses
- Magnetic resonance imaging [MRI] – principles and uses
- Nuclear radiation – types, dose, techniques.

B.1 X-rays

At GCSE level, you learned that X-rays are high-energy, ionising electromagnetic radiation. Their penetration depends on the concentration of electrons, so high density materials such as heavy metals can block X-rays, but low density materials such as flesh are relatively transparent to them. Because they are ionising, they are also mutagenic, i.e. can damage biological molecules, including DNA.

(a) The production of X-rays

This is a schematic diagram of an X-ray tube:

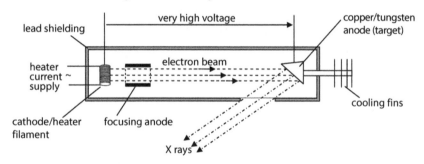

Fig. B1 X-ray tube

The heated cathode emits electrons by thermal emission; they are accelerated and focused by the focusing anode and then accelerated by a high voltage and hit a tungsten anode, where they are absorbed.

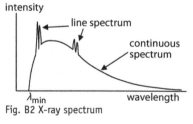

Fig. B2 X-ray spectrum

The rapid deceleration of the X-rays produces a continuous spectrum, the details of which depend upon the voltage.

Some electrons knock out inner electrons from the target atoms; electrons from higher energy levels drop into the empty energy levels, producing a line spectrum that is characteristic of the element.

Grade boost

Remember the basic method of X-ray production – accelerating electrons at a metal target.

Pointer

Increasing the heater current increases the intensity of the X-rays – more 'hot' electrons emitted and more X-rays produced.

Pointer

The photon energy is increased by increasing the accelerating pd.

Grade boost

The optimum photon energy for radiography is ~ 30 keV, achieved with a tube voltage of 50 – 100 keV.

Pointer

λ_{min} can be calculated from $\lambda_{min} = \dfrac{hc}{eV}$ but it's also common to have to rearrange the equation to obtain the accelerating pd (V).

① What is the maximum photon energy if the tube voltage is 60 kV?

(b) Controlling the X-ray beam and images

- Intensity – the higher the heater current, the hotter the cathode, so the more electrons are given off per second and the more X-ray photons are produced.

- Photon energy – the higher the accelerating voltage, the more energetic the electrons and the smaller the cut-off wavelength λ_{min}. The maximum photon

  energy $= eV$ (conservation of energy). Hence, $\dfrac{hc}{\lambda_{min}} = eV$.

- The image sharpness – placing a directional grid on the exit window increases sharpness (but cuts down the beam intensity).

- Contrast – selecting the accelerating voltage to produce a suitable range of wavelengths which are absorbed differently by the different materials, e.g. the softer the X-rays [lower energy] the better they will be absorbed by low density materials. Highly absorbent materials can be used to increase the contrast in soft tissue X-ray images, e.g. barium sulfate for alimentary canal imaging, or dyes containing iodine to make the coronary blood vessels show up.

- 3D images – computerised axial tomography [CT] scans use rotating beams which move along the body to produce images from all directions, from which 3D images can be constructed.

quickpire

② Calculate the cut-off wavelength λ_{min} if the tube voltage is 50 keV.

quickpire

③ The wavelength of the K_β line in the X-ray spectrum of copper is 0.14 nm. Estimate the minimum tube voltage needed to produce this wavelength and explain why this tube voltage is not quite enough to produce a line spectrum at this wavelength.

X-ray attenuation (absorption)

When X-rays pass through a material, the intensity, I, of the beam falls with distance, x, according to the equation

$$I = I_0 e^{-\mu x}$$

where μ is a constant that is characteristic of the material, called the attenuation [or absorption] coefficient.

Example

The attenuation coefficient of a material for an X-ray beam is 0.30 m^{-1}. What fraction of the incident photons remains after 5.0 m?

Answer

$$I = I_0 e^{-\mu x} = I_0 e^{-0.30 \times 5.0} = 0.22 I_0$$

So the fraction remaining $= 0.22$, ie. 22%.

Example

Find an expression for the 'half value thickness', $x_{\frac{1}{2}}$, the thickness of a material which reduces the intensity of X-rays by half.

Answer

Starting from the exponential decay of intensity with thickness: $I = I_0 e^{-\mu x}$, when $x = x_{\frac{1}{2}}$ we know that the intensity drops to $I = \frac{1}{2} I_0$

≫ Pointer

In the expression $e^{-\mu x}$, the product μx must have no units. It doesn't matter what the units of μ and x are as long as they are compatible, e.g.
$[x] = \text{cm}$ $\qquad [\mu] = \text{cm}^{-1}$
or
$[x] = \text{m}$ $\qquad [\mu] = \text{m}^{-1}$

quickpire

④ Calculate the distance for X-rays to halve their intensity if $\mu = 10 \text{ cm}^{-1}$.

[Hint: keep the unit as cm^{-1}]

Grade boost

Remember how to derive the equation $\mu x_{\frac{1}{2}} = \ln 2$ because the half value thickness often rears its ugly head when X-ray absorption is involved.

MATHS TIP

Remember that $\ln(e^a) = a$. See the Maths chapter for more practice with logs and exponentials.

quickfire

⑤ If the absorption coefficient $\mu = 0.3\,\text{cm}^{-1}$, calculate the thickness required to reduce the intensity of X-rays by 70%.

quickfire

⑥ A red scintillator crystal is 22% efficient and absorbs an X-ray photon of energy $250\,\text{keV}$. How many visible (red) photons of energy $2\,\text{eV}$ will the scintillator emit for each X-ray photon?

» Pointer

Note that the hand and image in Fig. B3 have been rotated for clarity.

» Pointer

The second scintillator plate (Fig. B3) increases the image brightness by a factor of 2 because only a small fraction of the X-rays are absorbed by the first scintillator and photographic plate.

---

giving $\frac{1}{2}I_0 = I_0 e^{-\mu x_{1/2}} \rightarrow \frac{1}{2} = e^{-\mu x_{1/2}} \rightarrow e^{\mu x_{1/2}} = 2$ taking (natural) logs leads to a very similar equation to that involving half-life and decay constant

$$x_{\frac{1}{2}} = \frac{\ln 2}{\mu} \quad \text{compared with} \quad T_{\frac{1}{2}} = \frac{\ln 2}{\lambda} \quad \text{for nuclear decay}$$

(d) Image intensification of X-rays (and γ-rays)

Image intensification has led to a massive decrease in the dosage received by patients for an average X-ray scan. All methods of image intensification are based on the following:

- X-ray photon energy is approximately $30\,000\,\text{eV}$ ($30\,\text{keV}$).
- Photon energy of light is approximately $3\,\text{eV}$.
- So, in theory, one X-ray photon (of energy $30\,\text{keV}$) can produce $10\,000$ light photons.

Devices called **scintillators** are designed using materials that are luminescent (or fluorescent) – they absorb higher energy photons and re-emit their energy as visible photons. In practice, the most commonly used scintillator material, sodium iodide (NaI), can emit up to 40 light photons for every $1000\,\text{eV}$ of incident X-rays ($\sim 10\%$ efficiency). So, a NaI scintillator can produce around 1000 light photons from one $30\,\text{keV}$ X-ray photon.

Whereas one X-ray photon will not have much effect on a photographic film, 1000 light photons will have a considerable effect. This means that a photographic film X-ray scan can be developed with a far lower dosage of X-rays than was previously possible without scintillators (the dosage of X-ray is decreased by a factor of about 1000).

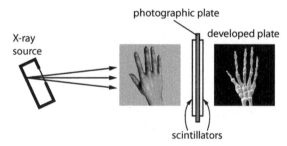

Fig. B3 Image enhancement using scintillators

The simplest way of using scintillators for image enhancement is shown in Fig. B3 and functions as follows:

The photographic plate is sandwiched between two large scintillator plates (this is called an X-ray cassette). Some X-ray photons are absorbed by the front scintillator and some by the back scintillator, these absorbed photons lead to around 1000 visible light photons. These light photons result in the photographic plate being exposed far quicker than without the scintillators. The dosage received by the patient is a tiny fraction of what it would be without the scintillators.

(e) Fluoroscopy

This is the process of viewing moving X-ray images in real time. It has been around for 120 years but recent developments mean that modern moving pictures are produced without lethal/harmful dosages of X-rays. Fig. B4 shows a possible arrangement which works as follows:

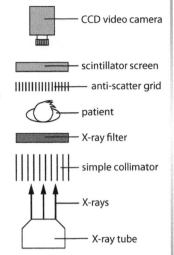

- The tube, simple collimator and filter just ensure that we have a nicely aligned X-ray beam of the correct energy (around $30\,\text{keV}$).

- The anti-scatter grid ensures that scattered X-rays from the patient not travelling in the correct direction are absorbed and not detected (similar to a collimator on a gamma camera but thinner and with a finer mesh).

- The scintillator converts each absorbed X-ray photon into around a thousand light photons (it's also housed inside a dark box for image clarity).

Fig. B4 X-ray fluoroscopy

- The CCD video camera takes regular images of the scintillator screen and these images are sent to a monitor that can be viewed by the medical team.

>> **Pointer**

The CCD video camera in fluoroscopy is an example of a digital image receptor but most X-ray images are still produced using photographic films (and image intensifiers).

(f) CT scanning

X-rays can also be used to obtain 3 dimensional images. This is usually done by completing a helical scan of the patient using a fan shaped X-ray beam with a detector opposite (see Fig. B5). The detector collects information continuously as both the source and detector are rotated around the patient. A computer is required to analyse all this information and come up with a 'best fit' 3D image of the patient. Excellent 3D images can be viewed on a popular file-sharing website (try a search on 'whole body CT scan').

A CT scan has two major advantages over conventional X-ray images:

1. A 3D image is produced.
2. Better soft tissue contrast is achieved.

However, there are also disadvantages:

1. Increased X-ray dosage.
2. Increased cost and time – usually a preliminary scan is followed by detailed scans.

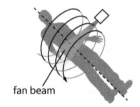

fan beam

Fig. B5 CT scan

>> **Pointer**

Another example of the use of a digital image receptor for X-rays is in CT scanning.

(g) Radiotherapy

Although X-rays can cause cancer, they can also be used to kill cancerous cells because cancerous cells are more readily killed than healthy cells.

Whereas X-rays for imaging require a high contrast between bone and soft tissue, the opposite is required for radiotherapy. X-rays with lower contrast and absorption are required so that all tissue types can be radiated and so that any depth of tissue can be radiated. This means that higher energy X-rays (or even gamma rays) are required and the energy range 1 MeV to 25 MeV is usually employed.

Fig. B6 Radiotherapy in action

The dosage to healthy tissue is reduced by the following steps:

- The beam is rotated about the tumour so that the X-rays are effectively 'focused' on the tumour (see Fig. B6).
- The correct photon energy is chosen so that the beam penetrates to the appropriate depth (lower energy for skin, higher energy for deep tumours).
- The beam is collimated to the correct width for the size of tumour.
- Masking is provided for the patient for the parts not requiring irradiation.

B.2 Radiation dosage

Absorption of ionising radiation leads to unwanted chemical reactions due to the ions produced. Small doses lead to no ill effects whatsoever but large dosages can lead to death. It is important to differentiate between these levels of dosage with good physical units.

There are three important definitions and two units of dosage that you must learn.

- Absorbed dose (D), measured in gray (Gy).
- Equivalent dose (H), measured in sievert (Sv).
- Effective dose (E), also measured in sievert (Sv).

The **absorbed dose**, D, is simply the energy of ionising radiation absorbed per unit mass. Hence, the unit Gy could be replaced by J kg^{-1}.

The **equivalent dose**, H, takes into account the relative danger of the type of ionising radiation itself. It is defined by:

$$H = DW_R$$

where W_R is the weighting factor of the radiation. Alpha is far more ionising than beta or gamma radiation and hence alpha has a higher weighting factor ($W_R = 20$) whereas both beta and gamma radiation have a weighting factor of 1.

The **effective dose** (E) takes into account both the relative danger of the ionising radiation and the susceptibility of the tissue to radiation damage. It is defined by:

$$E = HW_T$$

i.e. you take the equivalent dose and then multiply it by a further factor that depends on the tissue absorbing the radiation. W_T, it will not surprise you, is called the tissue weighting factor and varies from 0.01 for skin, brain and bone surface to 0.12 for lungs, colon and red bone marrow.

Example

The equivalent dose for skin irradiated by alpha particles is 22 mSv. Calculate the effective dose and the absorbed dose.

Answer

From above: for α, $W_R = 20$ and for skin $W_T = 0.01$.

$\therefore$ Effective dose, $E = HW_T = 22 \times 0.01 = 0.22$ mSv

and rearranging the equation for the equivalent dose gives:

$$\text{Equivalent dose, } H = DW_R \quad \rightarrow \quad D = \frac{H}{W_R} = \frac{22}{20} = 1.1 \text{ mGy}$$

Trickier example

An X-ray machine used for radiotherapy has an input power of 12 MW. It produces a beam of X-rays of intensity 18 W m^{-2} over an area of 5.0 cm × 5.0 cm. This beam irradiates a brain tumour which lies 6.2 cm below the surface of the skin. The half-value thickness of the skull/brain for these X-rays is 14 cm. Calculate:

(a) The efficiency of the X-ray machine.

(b) The absorption coefficient for the skull/brain.

(c) The time required for the tumour to receive an effective dose of 1.5 Sv if the tumour has a mass of 73 g and absorbs 6% of the X-rays incident upon it.

Answers

(a) Using the definition of intensity

$P = I \times \text{Area} = 18 \text{ W m}^{-2} \times (0.05 \text{ m})^2 = 0.045 \text{ W}$

hence,

$\text{Efficiency} = \dfrac{0.045 \text{ W}}{12 \times 10^6 \text{ W}} \times 100\% = 3.75 \times 10^{-7}\%$

(b) Using the equation derived earlier – Section B.1(c)

$\mu = \dfrac{\ln 2}{x_{\frac{1}{2}}} = \dfrac{\ln 2}{14 \text{ cm}} = 0.0495 \text{ cm}^{-1}$

(c) First, use the absorption coefficient to obtain the intensity at the tumour: $I = I_0 e^{-\mu x}$

And use the links between power and intensity and between power and energy:

$\text{Energy} = Pt \quad \text{and} \quad P = IA \quad \rightarrow \quad \text{Energy} = IAt$

Combining these two: $\qquad \text{Energy} = AtI_0 e^{-\mu x}$

---

quickfire

⑦ The absorbed dose for kidneys irradiated by gamma rays is 1.2 mJ g^{-1}. Calculate the equivalent and effective doses.

$W_R = 1$ and $W_T = 0.05$

quickfire

⑧ The effective dose for gonads irradiated by alpha particles is 0.7 mJ g^{-1}. Calculate the equivalent and absorbed doses.

$W_R = 20$ and $W_T = 0.08$

Grade boost

There is no need to convert from mSv to Sv.

Grade boost

Don't worry too much if the values of weighting factor on the exam paper are different from the ones you've seen before, there seems to be disagreement among the experts as to the best values.

Pointer

Part (c) is tough and probably too long to include on an exam paper – it involves 7 equations and a lot of substitution and rearranging. This would probably be worth around 8 marks. However, remember the individual steps, a watered down version would be an appropriate question for around 4 marks.

Grade boost

When you have this many equations to juggle to obtain the correct answer, the Data booklet is not much use. **You need a good grasp and understanding of the equations without the Data booklet.**

Now, we must use the definitions of absorbed, equivalent and effective dose:

$$D = \frac{\text{Energy}}{m}, \text{ plus } H = DW_R, \text{ plus } E = HW_T \rightarrow E = \frac{\text{Energy} \times W_R W_T}{m}$$

However, the energy absorbed is only 6% of the incident energy:

$$\therefore E = \frac{0.06 \times \text{Energy} \times W_R W_T}{m} = \frac{0.06 \times AtI_0 e^{-\mu x} W_R W_T}{m}$$

Rearrangement followed by substitution of all the relevant values gives:

$$t = \frac{Em}{0.06 \times AtI_0 e^{-\mu x} W_R W_T} = \frac{1.5 \times 73 \times 10^{-3}}{0.06 \times (0.05)^2 \times 18 e^{-0.0495 \times 62} \times 1 \times 0.01}$$

$$= 5\,500 \text{ s}$$

This value assumes that the tumour has the exact dimensions of $5\,cm \times 5\,cm$ and that it all receives the same intensity of radiation. However, this is the only calculation possible from the information provided.

» Pointer

The compound eyes of insects work on the same principle as the collimator in a gamma camera. They have no lenses but only allow light rays to hit the retina which are travelling parallel to the axis of each tiny eye. The brain then constructs the image. The praying mantis even manages 3D images using these eyes.

B.3 Gamma camera

Gamma cameras are used regularly in hospitals. Its principles are illustrated in Fig. B7.

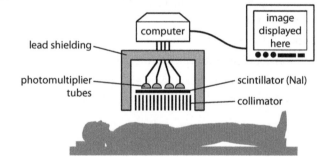

Fig. B7 Principle of operation of a gamma camera

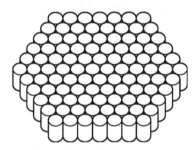

Fig. B8 Lead collimator tubes

» Pointer

In gamma cameras, photomultipliers are gradually being superseded by CCD cameras which are far cheaper and the gamma camera set-up is becoming more similar to that used for fluoroscopy.

There is no way of focusing γ-rays, so a lead collimator is used to ensure that only the γ rays travelling vertically upwards (in Fig. B7) hit the crystal. The lead collimator is just a 2D arrangement of lead tubes that absorb gamma rays not travelling parallel to their axes. Because these collimators have to be reasonably thick to absorb the gamma rays, the resolution of gamma camera images is not very high (the hexagonal collimator illustrated in Fig. B8 would have a side of length $5\,cm$ so the separation of pixels is around $5\,mm$). After the collimator, there is a scintillator (**NaI** again) that converts a gamma ray to thousands of visible photons. The photomultiplier tubes detect the light flashes. The computer and electronics process and average this information to produce 2D images.

B.4 Gamma-emitting radio tracers

Radio tracers are chemical compounds in which one of the atoms has been replaced by a radioactive isotope. Such tracers have a wide industrial application, e.g. finding the location of cracks in buried gas pipelines. They are also used in medicine for locating medical problems and for imaging. One of the most common radiotracers used is the gamma emitting isotope of technetium, Tc99m. It has a half-life of only 30 minutes, so hospitals have to produce it on site.

The m in nuclide Tc99m stands for 'metastable' and indicates that this nuclide is in an excited state. It decays to Tc99 with the emission of a 0.14 MeV photon (γ), which can be used for imaging. The patient is given an injection of a compound containing the Tc99m (often $Na^+ TcO_4^-$). The compound is chosen to be readily absorbed by tumours, so images of the patient taken using a gamma camera (see above) will reveal the presence of tumours. Tc99m is frequently used in heart and brain imaging.

Grade boost

The main properties of a radioactive tracer are:
- It behaves chemically as required, i.e. it goes where required without being poisonous/harmful.
- It emits the correct type of radiation (usually gamma).
- It has a half-life of the order of an hour – long enough to do the experiment but short enough to decay in a reasonable time without a large dose to the patient.

B.5 Positron emission tomography (PET scan)

In PET scans, a glucose molecule is usually tagged with a positron-emitting atom, such as fluorine-18. The two gamma photons produced when the e^+ and an electron mutually annihilate can be detected outside the body using a gamma camera.

Cancers contain actively dividing cells and so have a high requirement for energy and tend to concentrate glucose. Hence PET scans are useful for cancer imaging.

quickpire

⑨ A patient is given a gamma-emitting radio tracer with a half-life of 6 hours and is told not to go near pregnant women or children for 3 days (for the activity to drop to a safe level). Another patient is given 4× the dose of the same gamma emitter to obtain a more detailed gamma camera image. How long must the second patient stay away from pregnant women and children?

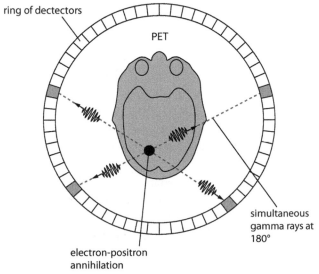

Fig. B9 The principle of the PET scan

In a PET scanner (see Fig. B9), a circular gamma camera surrounds the patient. When a positron annihilates an electron the two emitted gamma rays are detected almost simultaneously. The source of the two gamma rays must then lie on a line connecting the points where the gamma rays were detected.

The gamma camera can measure time delays of around 10^{-12} s and from this the position of the source of the gamma rays can be located along the line between the two detectors, e.g. if the pulse reaches the left detector 5 ps before it reaches the right detector then the source is

$$5 \times 10^{-12} \times 3 \times 10^8 = 15 \times 10^{-4}\,\text{m}$$

nearer to the left detector than the right (because gamma rays travel at the speed of light). Using this information a 3D image of positron annihilation hotspots can be constructed.

quickfire

⑩ Calculate the accuracy of timing required for a PET scan to have a spatial resolution of 0.5 mm.

Key Terms

A-scan = a 1D scan where time delays give distances and thicknesses.

B-scan = a 2D scan (usually moving) giving low resolution images of organs/foetuses.

Acoustic impedance = $Z = c\rho$ where c is the speed of sound and ρ is the density.

Coupling medium = a gel that matches the acoustic impedance of the two materials at a boundary, it also eliminates air from the boundary.

B.6 Ultrasound

This is produced using a piezoelectric crystal. These crystals deform in response to an electric field. Electrodes apply a high frequency alternating pd to the crystal which vibrates at the same frequency and generates the sound wave. The process also works in reverse: the crystal produces an alternating pd in response to an incident sound wave – thus the transmitter is also a detector. This is a typical ultrasound probe.

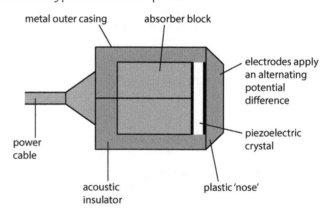

Fig. B10 Ultrasound probe

The absorber block soaks up the sound waves which would be sent to the left (in Fig. B10). These would be subsequently reflected and interfere with the reflected signals from the body.

A pulse of high frequency [MHz] alternating pd is applied to the piezoelectric crystal; a pulse of sound waves is produced and travels to the right into the body; reflections are received [see below]; the crystal converts these into electrical pulses which are sent back along the power cable and analysed. This is repeated many times per second.

Types of scan	Description	Example
A-scan	1 dimensional; return pulses detected on CRO; time delay used for determining distances or existence of structures.	A tumour would alter the time for a reflected pulse from a known structure.
B-scan	2 dimensional: an array of detectors or a single moving transmitter/detector used; return pulses displayed on screen; images of structures seen.	Foetal / prenatal scans to give images and information about the size and development of a foetus (also the number of foetuses in sheep is important).

quicKpire

⑪ Show that the unit of Z is $kg\,m^{-2}\,s^{-1}$.

Key Terms

Piezoelectric effect – a material produces an electric field (or pd) when deformed.

Reverse piezoelectric effect = a material deforms when subjected to an electric field (or pd).

(a) The reflection of ultrasound

Ultrasound is reflected whenever it crosses a boundary between two media. The property of the materials which determines the fraction reflected is the **acoustic impedance**, Z, of each of the media. This is defined by:

$$Z = c\rho$$

where c is the speed of sound (in the medium) and ρ the density of the medium. Some figures for reference:

$$Z_{air} \sim 400\,kg\,m^{-2}\,s^{-1} \quad \text{and} \quad Z_{skin} \sim 2 \times 10^6\,kg\,m^{-2}\,s^{-1}.$$

The fraction, R, of the ultrasound energy reflected is given by

$$R = \frac{(Z_2 - Z_1)^2}{(Z_2 + Z_1)^2}$$

If you need it in the examination you will be given this equation in the question paper (it is not in the Data booklet). Using the above values of Z for air and skin, you should be able to show that R between air and skin is almost 100%, i.e. hardly any ultrasound (<0.1%) would penetrate into the body from air, and hardly any sound from within the body would emerge back into the air. There would always be a thin layer of air between the ultrasound probe and dry skin, so a **coupling medium**, a gel, is applied to the skin and the probe placed in contact with the gel. The gel has a value of Z which is almost the same as skin so hardly any ultrasound is reflected at the gel–skin boundary – it all gets across in both directions.

≫ Pointer
The fraction of ultrasound reflected at a boundary
$R = \dfrac{(Z_2 - Z_1)^2}{(Z_2 + Z_1)^2}$ is for
ultrasound incident along the normal to the interface. Non-normal incidence will never be examined because the equations are well beyond the scope of A-level physics – and most physics degrees!

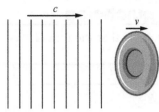

Fig. B11 Sound wave hitting a red blood cell

quickpire

⑫ What fraction of sound is reflected at a boundary between fresh water and sea water?

$Z_{fresh} = 1.43 \times 10^6 \, kg\,m^{-2}s^{-1}$;

$Z_{sea} = 1.45 \times 10^6 \, kg\,m^{-2}s^{-1}$.

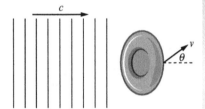

Fig. B12 Sound wave hitting at an angle

quickpire

⑬ In practice, the angle θ might not be known. Does this give a significant error if θ is less than 15° and an accuracy of 10% is required?

Grade boost

Remember that

$$\frac{\Delta\lambda}{\lambda} = \frac{\Delta f}{f}$$

(b) Doppler probe

This is an alternative way of using ultrasound to study the flow of blood. The technique is based on the same physics as the detection of extra-solar planets by looking at the star's spectrum.

Consider an ultrasound wave of wavelength λ, hitting a red blood cell moving with velocity v. From Section 2.8, the wavelength of the sound wave which the red blood cell 'sees' is shifted by $\Delta\lambda$, given by the equation:

$$\frac{\Delta\lambda}{\lambda} = \frac{v}{c},$$

Also, the red blood cell reflects the ultrasound wave back to the detector along its original direction. Because the red blood cell is moving away from the detector, the wave is Doppler shifted for a second time by the same amount. So the received ultrasound signal has a value of $\Delta\lambda$, given by:

$$\frac{\Delta\lambda}{\lambda} = 2\frac{v}{c}$$

Usually the direction of motion of the red blood cell is not in exactly the same direction as the direction of propagation of the ultrasound (Fig. B12). In this case, we need the component of the red blood cell's velocity in the direction of the ultrasound ($v\cos\theta$), so the equation becomes:

$$\frac{\Delta\lambda}{\lambda} = \frac{2v\cos\theta}{c}$$

The fractional change in the wavelength is exactly the same as the fractional change in the frequency (except that one increases as the other decreases). The equation that appears in the specification and the data booklet is:

$$\frac{\Delta f}{f} = \frac{2v\cos\theta}{c}$$

Example

Ultrasound from a probe enters the aorta of a foetus at an angle of 25° to the direction of blood flow. The change in frequency of the reflected ultrasound is 1.3 kHz and the incident ultrasound has a frequency of 3.5 MHz. If the speed of the ultrasound waves is 1600 m s^{-1}, calculate the speed of flow of blood in the aorta of the foetus.

Answer

Rearranging for v gives:

$$v = \frac{c\Delta f}{2f\cos\theta} = \frac{1600 \times 1300}{2 \times 3.5 \times 10^6 \times \cos25°} = 0.33 \text{ m s}^{-1}$$

Believe it or not, this is a calculation performed automatically by ultrasound machines for unborn babies on a regular basis.

B.7 Magnetic resonance imaging (MRI)

This is a completely harmless technique that involves no ionising radiation and which produces highly detailed images especially of soft tissue. Unfortunately, the technique uses superconducting magnets and so cannot be used on patients with pacemakers (although non-ferrous, MRI-friendly pacemakers have been available since 2010).

It involves lying still for a long time (Fig. B13) and having radio waves fired at you. This causes problems to young children and is claustrophobic but this too can be overcome by putting the patient to sleep.

How does it work?

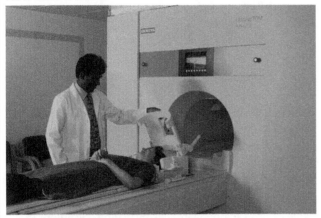

Fig. B13 MRI scan preparation (Science Photo Library)

(a) Magnetic nuclei

All nuclei spin but medical MRI scanning applies mainly to hydrogen nuclei (a proton) because the body contains 75% water (H_2O). In the absence of a magnetic field, the direction of the spin of hydrogen nuclei is random. When a magnetic field is applied, the nuclei line up with the magnetic field lines (Fig. B14). They, in fact, precess about the applied magnetic field similar to the way that a gyroscope

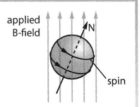

Fig. B14 Spinning nucleus 1

precesses about the applied gravitational field. The frequency at which the hydrogen nuclei spin about the applied magnetic field (B) is a very precise frequency and is given by the equation:

$$f = 42.6 \times 10^6 \, B$$

The **Larmor frequency** = the photon frequency which resonates with the energy level difference between the magnetic moment states of a nucleus in a magnetic field.

The **relaxation time** = a type of average time which nuclei take to flip their spins back in alignment with the applied B-field.
It varies greatly for different soft tissue.

quickfire

⑭ The B-field (in T) of a MRI machine is given by

$$B = 3.2 + 0.25x$$

where x is the distance in metre along the MRI table. Find the value of x where radio waves of frequency 137.9 MHz will be able to scan.

where B is the magnetic flux density measured in tesla (T). This is how the equation appears in the Data booklet and you will need to be able to calculate **Larmor frequencies** (or B-fields) with it. As this is the natural frequency of the nuclei, this is also the resonance frequency for forced oscillations and these oscillations can be forced by sending radio waves of this frequency at the hydrogen nuclei (done by sending current at this frequency to emitting coils).

When this is carried out, a large number of hydrogen nuclei absorb the radio waves and flip their alignment (see Fig. B15). This is a slightly higher energy state and the nuclei will remain in that state for a time before flipping back and emitting radio waves.

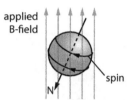

Fig. B15 Spinning nucleus 2

The characteristic time before the nuclei flip back (after the applied radio waves are switched off) is called the '**relaxation time**'. The emitted radio waves are detected by coils (often the same as the emitting coils). This 'relaxation time' is highly dependent on the concentration of hydrogen nuclei (water has a relaxation time of 2.5 s and body fat 0.18 s). Hence, MRI is capable of producing images with superb contrast of soft tissue.

An MRI scanner is capable of producing 3D images by having a gradient in the magnetic field (Fig. B.16).

The B-field half way along the scan is 1.5 T and the resonant frequency for $B = 1.5$ T is

$$f = 42.6 \text{ MHz T}^{-1} \times 1.5 \text{ T} = 63.9 \text{ MHz}$$

When radio waves of frequency 63.9 MHz are emitted, it is the slice shown in the diagram that will be scanned. By changing the frequency of the radio waves, different slices can be scanned.

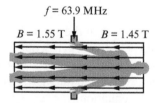

Fig. B16 Magnetic field gradient

B.8 Comparison table of the main imaging techniques

Property	Ultrasound	Standard X-ray	CT scan	MRI
Radiation exposure	No ionising radiation	Exposure to ionising radiation	Significant exposure to ionising radiation (2–10 mSv) up to 5 years background	No ionising radiation
Application	Generally soft tissue including foetus; skeletal joints	Mainly bone breakages, with contrast agent can also be used for soft tissue	Bone injuries, lung and chest imaging, cancer detection, A&E investigations	Different kinds of soft tissue imaging, e.g. injuries, tumours
Biological effects	No known hazards in imaging	Carcinogenic effects and developmental defects in embryos	As X-ray	No known hazards Allergic reaction to contrast agents
Cost	Low cost	Low cost	About half the cost of MRI	High cost
Conditions	Short time; relatively painless (probe, e.g. rectal, may be required)	Very short time	Quite short time (5 min), ideally no movement but less of a problem than MRI	Long time; uncomfortable (no movement allowed); noisy; claustrophobia
3D imaging	Not usually – each image is a 2D slice	Not without moving the patient	Possible using a helical scan	Yes
Definition	Not high – depends on the skill of practitioner	High definition of bony structures	High definition of bony structures; moderate definition of soft structures (better with contrast agent)	High definition (but requires stationary subject)
When not to use	When air gets in the way, e.g. lungs, air-filled intestines	Pregnancy	Weight limit of ~200 kg because of space and strength of moving table	Some metal implants; heart pacemaker Weight limit ~ 150 kg (space / strength of table)

1. The tube current of an X-ray machine is 1.2 A and the accelerating pd is 100 kV.
   (a) Calculate the input power of the X-ray machine.
   (b) The total X-ray output of the machine is 50 W. Calculate the efficiency of the X-ray machine.
   (c) Calculate the maximum energy of X-ray photons in eV and J.
   (d) Calculate the minimum wavelength (λ_{min}).
   (e) Estimate the number of X-ray photons emitted per second.

2. A beam of $30\,\text{keV}$ X-ray photons has an attenuation coefficient of $0.9\,\text{cm}^{-1}$ in tissue and $9.0\,\text{cm}^{-1}$ in bone. In obtaining an X-ray image, part of the beam goes through $5\,\text{cm}$ of tissue and another part of the beam passes through $3\,\text{cm}$ of tissue and $2\,\text{cm}$ of bone.

   (a) Calculate the ratio of the intensities of the two parts of the beam. (Hint: the initial intensity I_0 will cancel out and, if you want to be lazy, both parts of the beam travel through $3\,\text{cm}$ of tissue.)

   (b) The number of photons arriving at the photographic plate per second and per mm^2 is 2.1×10^5 after passing through $5\,\text{cm}$ of tissue.

   (i) Calculate the number of photons per second per mm^2 arriving at the photographic plate having passed through $3\,\text{cm}$ of tissue and $2\,\text{cm}$ of bone.

   (ii) Calculate the intensity of the X-ray beam corresponding to 2.1×10^5 photons per second per mm^2.

3. An X-ray machine used for radiotherapy has an input power of $8\,\text{MW}$. It produces a beam of X-rays of intensity $15\,\text{W}\,\text{m}^{-2}$ over an area of $3.2\,\text{cm} \times 2.4\,\text{cm}$.

   (a) Calculate the efficiency of the radiotherapy machine for these settings.

   (b) This beam irradiates a lung tumour which lies $8.5\,\text{cm}$ below the surface of the skin. The half value thickness of the ribs/lung tissue for these X-rays is $11.6\,\text{cm}$. Calculate:

   (i) The absorption coefficient for the ribs/lung tissue.

   (ii) The intensity of the beam at the depth of the tumour.

   (iii) The absorbed dose per minute for the tumour if the tumour has a mass of $220\,\text{g}$ and absorbs 20% of the radiation incident upon it.

   (iv) State why the absorbed dose is also the equivalent dose.

   (v) Calculate the time taken for the effective dose of the lung tumour to reach $3.0\,\text{Sv}$. [$W_T = 0.12$ for lungs]

4. (a) Use the data given in the table to explain why gel (coupling material) is used in ultrasonic imaging.

	Plastic nose of ultrasound probe	Air	Gel	Skin
Speed of sound (m s^{-1})	1800	340	1595	1500
density (kg m^{-3})	940	1.25	960	1050

   (b) The speed of sound in blood is $1570\,\text{m s}^{-1}$. Calculate the change in frequency expected for ultrasound reflected from red blood cells travelling at $0.45\,\text{m s}^{-1}$ when the initial frequency of the ultrasound is $7.5\,\text{MHz}$. State any assumption that you make.

5. (a) An MRI machine has a B-field that varies from $8.0\,\text{T}$ at one end of the patient to $7.5\,\text{T}$ at the other end. Calculate the range of the Larmor frequencies used for this varying magnetic field.

   (b) List the advantages and disadvantages of a MRI scan over an ultrasound scan.

   (c) List the advantages and disadvantages of a MRI brain scan over PET brain scan.

6. State why PET, MRI, CT, ultrasound or standard X-ray scans would or would not be used in the following instances:

   (a) A scan of a new born baby's brain.

   (b) A scan of an adult's brain.

   (c) A lung scan to search for cancer.

   (d) A lung scan to search for fluid.

   (e) A scan to find the location of a broken bone in the foot.

Option C: The physics of sports

We've already met most of the physics principles in earlier sections of the Study and Revision Guides (SRG). The physics of rotational motion and Bernoulli's equation are now introduced. You should be prepared to apply all these principles in whatever sporting context the examiner chooses to set questions!

C.1 Moments and stability

This part of the option relates to the material developed in Sections 1.1.4 and 1.1.5 of the AS SRG. As we shall see, it also has a strong connection to the concept of Centre of mass in Section 2.8 of the specification. For many purposes we can treat the body as if its weight were concentrated at one single point – called the **centre of gravity** (CoG).

C.1.1 Finding the centre of gravity

(a) For symmetrical objects

See Fig. 1.1.13 in the AS SRG for the location of the CoG of symmetrical objects.

(b) Using the definition

For objects without intersecting planes of symmetry we can calculate the position of the centre of gravity, **C**, by taking moments. If we imagine all the weight of the system to be concentrated at **C**, its moment about any point must be the same as the sum of the moments of the different parts of the object.

Example

Find the position of the CoG, **C**, of the system of the three masses which are held at the corners of of an equilateral triangle by light (i.e. negligible mass) rods in Fig. C1

Answer

There is one plane of symmetry (the pecked line) so C must lie on it. Let it be a distance x from the base of the triangle as shown. We'll assume the masses are arranged in a horizontal plane.

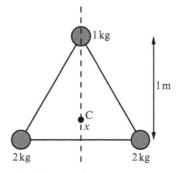

Fig. C1 System of masses

Taking moments about the base line, the only mass with a moment is the 1 kg one. Its moment, M, is given by

$$M = 1\,\text{kg} \times 9.81\,\text{N kg}^{-1} \times 1\,\text{m} = 9.81\,\text{N m}$$

> ## Pointer
>
> If a body has *reflectional symmetry*, the CoG must lie on a plane of symmetry (a line of symmetry for a plane figure). If a body has several planes of symmetry, the CoG must be at their point of intersection.

quickfire

① Find the value of x in Fig. C1 by taking moments about a line through C parallel to the base of the triangle.

The total weight of the system is $(5 \text{ kg}) \times 9.81 \text{ N kg}^{-1} = 49.05 \text{ N}$. If all this weight were concentrated at **C** its moment about the base line would be $49.05x$ N.

These two moments must be equal, i.e. $49.05x \text{ N} = 9.81 \text{ N m}$

$$x = \frac{9.81 \text{ N m}}{49.05 \text{ N}} = 0.2 \text{ m}$$

(c) Using the balance point

The system will balance if we put it onto a fulcrum at its centre of mass because the clockwise and anticlockwise moments will be equal.

Example

Find the position of the balance point (i.e. the centre of mass) of the system of two masses in Fig. C2.

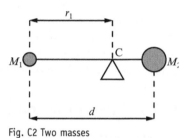

Fig. C2 Two masses

Answer

Taking moments about **C**:

For equilibrium: $M_1gr_1 = M_2g(d - r_1)$

Dividing by g and rearranging $(M_1 + M_2)r_1 = M_2d$

Hence, dividing by $M_1 + M_2$ $r_1 = \frac{M_2}{M_1 + M_2}d$

Note that this equation is the same as that for the **Centre of mass** in the Orbits section of the specification. This is no accident: if the objects are in a uniform gravitational field the CoM and CoG coincide.

C.1.2 Stability

Here we are concerned with whether an object or a person will fall over or capsize when it is pushed from the side. The general rule is:

Wide base; low centre of gravity   Stable

Narrow base; high centre of gravity Unstable

The box in Fig. C3 has a weight (mg) of 600 N. Making a reasonable assumption about its symmetry you should be able to show that the minimum force needed to make the box start to tip is 250 N. And the lower the centre of gravity, C, the greater the angle the box has to turn through before it tips (see Quickfires 2 and 3).

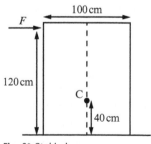

Fig. C3 Stable box

quickfire

② Show that the minimum force needed to tip the box in Fig. C3 is 250 N.

Example

Calculate the angle through which the box in Fig. C3 needs to be turned before it tips over.

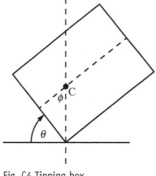

Fig. C4 Tipping box

Answer

Fig. C4 shows the box when it is on the point of tipping, i.e. when the centre of gravity is vertically above the pivot point. It has turned through angle θ. Using the properties of angles in a triangle, it is clear (see Quickfire 3) that the marked angle at C is also θ.

$$\tan\theta = \frac{\text{opp}}{\text{adj}} = \frac{50 \text{ cm}}{40 \text{ cm}} = 1.25. \quad \therefore \quad \theta = \tan^{-1} 1.25 = 51.3°$$

quickfire

③ Show that the angles ϕ and θ in Fig. C4 are the same.

C.2 Forces and collisions

The basic facts of this were developed in Section 1.3 especially 1.3.4 and 1.3.6 of the AS SRG.

C.2.1 Newton's second law of motion (N2)

N2 states that the mean resultant force on a body is given (in SI) by $F = \dfrac{\Delta p}{t}$, where p is the momentum of the body.

Hence, with the usual symbols, $F = \dfrac{mv - mu}{t}$

Hence $Ft = mv - mu$

which is given as an equation in the Data booklet.

This form of N2 is useful in many sporting contexts. For example, consider Fig. C5, which shows water flow being diverted by the rudder of a sailing dinghy (seen from the frame of reference of the boat – so the water is moving and the rudder is stationary). The example shows how we can use $Ft = mv - mu$ to calculate the force exerted by the water on the rudder – using N2 and N3.

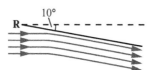

Fig. C5 Diverted water flow

Key Term

Newton's second law of motion states that the rate of change of momentum of a body is directly proportional to the resultant force acting on it. In SI units, the constant of proportionality is defined as 1, so

$$F = \frac{\Delta p}{t}$$

Example

The rudder, **R**, in Fig. C5 is in a stream of water flowing at $2.5\,\text{m s}^{-1}$. It diverts the flow of $0.25\,\text{m}^3$ of water through $10°$ each second. Calculate the force exerted by the water on the rudder.

Answer

Mass of water diverted each second = volume each second × density

$$= 0.25\,\text{m}^3 \times 1000\,\text{kg m}^{-3}$$

$$= 250\,\text{kg}$$

From the vector diagram, Fig. C6, we can show that the change in velocity of the water, $v - u$, is $0.44\,\text{m s}^{-1}$ at an angle of $5°$ to the vertical.

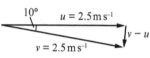

$\therefore$ The force on the water $= \dfrac{mv - mu}{t} = \dfrac{m(v - u)}{t}$

Fig. C6 Calculating Δv

$$= \frac{250\,\text{kg} \times 0.44\,\text{m s}^{-1}}{1\,\text{s}} = 110\,\text{N}$$

So the rudder exerts a force on the water to the side (and slightly forward) of $110\,\text{N}$ so, by N3, the water exerts a force of $110\,\text{N}$ in the opposite direction.

quickfire

④ Show that the vector $v - u$, in Fig. C6 is $0.44\,\text{m s}^{-1}$ at $5°$ to the vertical.

Hint: Isosceles triangle.

quickfire

⑤ In the example, what backward force (i.e. horizontally to the right) does the water exert on the rudder?

C.2.2 The coefficient of restitution, e

Section 1.3.6 in the AS SRG dealt with how to solve collisions problems using conservation of momentum in situations where kinetic energy is conserved (*elastic collisions*) and where it is not (*inelastic collisions*). The **coefficient of restitution**, e, can be used to simplify calculations. For a collision between two objects, it is defined by:

$$e = \frac{\text{relative speed after collision}}{\text{relative speed before collision}}$$

Key Terms

The **relative velocity** of body B to body A is defined as $v_B - v_A$, where v_B and v_A are the velocities of **B** and **A** respectively.

The **relative speed** of two objects is the magnitude of the relative velocity.

E.g. if $v_A = 10\,\text{m s}^{-1}$ and $v_B = 2\,\text{m s}^{-1}$, the velocity of **B** relative to A is $-8\,\text{m s}^{-1}$ and the relative speed of the two is $8\,\text{m s}^{-1}$.

(a) Colliding objects

In the following collision

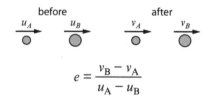

$$e = \frac{v_B - v_A}{u_A - u_B}$$

Fig. C7 Definition of coefficient of restitution

Example

A body, **A**, of mass $2\,\text{kg}$ travelling at $10\,\text{m s}^{-1}$ collides head on with a stationary body, **B**, of mass $3\,\text{kg}$. The coefficient of restitution is 0.8. Calculate the velocities of the two bodies after the collision.

quickfire

⑥ Show that substituting the value of v_B into equations [1] and [2] gives the same value for v_A and find this value.

Answer

Using the symbols in Fig. C7, with $u_A = 10\,\text{m s}^{-1}$ and $u_B = 0$

Conservation of momentum: $\qquad m_A u_A = m_A v_A + m_B v_B$

Substituting $\qquad \therefore\quad 20 = 2v_A + 3v_B \qquad [1]$

Coefficient of restitution $\qquad 0.8 = \dfrac{v_B - v_A}{u_A}$

Rearranging & substituting $\qquad \therefore\quad 8 = -v_A + v_B \qquad [2]$

Solving; $[1] + 2\times[2]$ $\qquad \rightarrow\quad 36 = (3 + 2)v_B$

$\qquad\qquad\qquad\qquad\qquad \therefore\quad v_B = 7.2\,\text{m s}^{-1}$

Substituting back into equations [1] or [2] gives us v_A (see Quickfire 6). Note that we have found the final velocities without having to solve quadratic equations.

(b) Bouncing balls

A bouncing object is a special case of colliding objects in which the mass of one of them (the ball) has a negligible mass compared to the other (the Earth). In principle, the bounce does impart some motion to the Earth but it is negligibly small. Hence the situation simplifies to:

$$e = \frac{v}{u}$$

Fig. C8 Coefficient of restitution – bouncing ball

If the ball is dropped at zero speed from a height H, it will hit the ground with a speed $\sqrt{2gH}$ (see Quickfire 9). Its bounce speed is therefore by definition $e\sqrt{2gH}$. We can use $v^2 = u^2 + 2ax$ to work out the bounce height, h as follows:

At height h the velocity is zero, $\qquad \therefore\ 0 = e^2 2gH - 2gh$

Dividing by 2g and rearranging gives $\qquad e^2 = \dfrac{h}{H},\ \therefore\ e = \sqrt{\dfrac{h}{H}}$

Example

The coefficient of restitution for a table tennis ball dropped onto a heavy steel plate is 0.790. Calculate the rebound height when the ball is dropped from a height of 1.50 m in a vacuum (see Quickfire 10).

Answer

$$e = \sqrt{\frac{h}{H}},\ \therefore\ 0.790 = \sqrt{\frac{h}{1.50\,\text{m}}}$$

Squaring and rearranging: $\rightarrow h = 0.790^2 \times 1.50\,\text{m} = 0.936\,\text{m}$

Note that the impacting surface doesn't have to be horizontal. See the **Extra questions** for an example of a tennis ball bouncing from a vertical wall.

quickfire

⑦ Calculate the fraction of the initial kinetic energy lost in the collision in the example. Does this mean that conservation of energy does not apply?

quickfire

⑧ Repeat the Example for the case when the 3 kg mass has an initial velocity of $-10\,\text{m s}^{-1}$. Find v_A and v_B.

≫ Pointer

In Fig. C8, the velocities u and v have opposite directions, so one would normally be shown as negative. The symbols in the equation refer to the speeds which are both positive.

quickfire

⑨ State why we can use the equation $v^2 = u^2 + 2ax$ to work out the impact speed and the bounce height.

Use this equation to show that the impact speed is $\sqrt{2gH}$.

Grade boost

The value of e is not strictly a constant. It generally decreases slowly with impact speed. For example, for a table tennis ball dropped from 1.70 m onto steel $e = 0.780$. Be prepared for an examiner to set a question with a variable value of e.

quickfire

⑩ Suggest a reason why the Example specifies 'in a vacuum'.

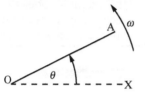

Fig. C9 Rotating rod

C.3 Rotational motion

Many sports involve the athlete rotating. Section 1.5 dealt with objects moving in a circle. In a sporting context, this could include a throwing hammer. In order to achieve the circular motion of the hammer, the thrower also has to accelerate him/herself, the understanding of which requires the development of additional concepts.

C.3.1 Kinematic quantities

These are the quantities we use to *describe* rotations, in the same way that we use displacement, velocity and acceleration to describe linear motion. The following table shows the relationship between the linear and rotational quantities.

Linear		Rotational	
Quantity	Symbol	Quantity	Symbol
Displacement	x	Angular position	θ
Velocity	v	Angular velocity	ω
Acceleration	a	Angular acceleration	α

Table C1 Comparison of linear and rotational quantities

You can easily write the rotational equations of motion for constant angular velocity or for constant acceleration, if you remember the equivalent linear equations. In addition to the quantities in Table C1 you just need to make the changes: $u \to \omega_1$; $v \to \omega_2$, e.g.

$$x = ut + \tfrac{1}{2}at^2 \quad \text{becomes} \quad \theta = \omega_1 t + \tfrac{1}{2}\alpha t^2$$

Example

A sling-shot thrower releases a stone at $30\,\text{m s}^{-1}$ from a 50 cm long sling after accelerating it from rest in 2 revolutions. Calculate the mean angular acceleration.

Answer

Angular velocity, at release $\omega_2 = \dfrac{v}{r} = \dfrac{30\,\text{m s}^{-1}}{0.50\,\text{m}} = 60\,\text{rad s}^{-1}$

Angle rotated, $\theta = 2 \times 2\pi = 4\pi$ rad

Remember $v^2 = u^2 + 2ax$

$\therefore \omega_2{}^2 = \omega_1{}^2 + 2\alpha\theta$

$\therefore 60^2 = 0 + 2\alpha \times 4\pi$

$\therefore \alpha = \dfrac{60^2}{8\pi} = 143$ rad s^{-2}

C.3.2 Rotational dynamics

This is the rotational equivalent of Newton's laws of (linear) motion. For example, N2 can be written:

$$F = ma \quad \text{or} \quad F = \frac{\Delta p}{t}$$

The rotational equivalents of these are

$$\tau = I\alpha \quad \text{and} \quad \tau = \frac{\Delta L}{t}$$

(a) Moment of inertia, *I*

This is the rotational equivalent of inertial mass. The more extended an object, the more difficult it is to start in rotating and the more kinetic energy it has when it is in motion. The definition in **Key terms** (which you should learn) looks quite complicated. A simple example (to which we'll keep returning) should clarify it.

Example

Calculate the moment of inertia of the two 5 kg spheres about the axis shown in Fig. C10. Ignore the mass of the connecting rod and take the spheres to be point masses.

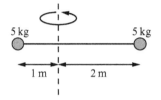

Fig. C10 Moment of inertia

Answer

$I = \Sigma m_i r_i^2 = 5\,\text{kg} \times (1\,\text{m})^2 + 5\,\text{kg} \times (2\,\text{m})^2 = 25\,\text{kg m}^2$

The moment of inertia depends upon the position of the axis of rotation. It is a minimum if the axis passes through the centre of mass (See Quickfire 12). Some formulae for moments of inertia are given in Table C2 but any such formula that you need will be given in an examination question.

Object	I	
Uniform sphere	$\frac{2}{5}mr^2$	Through the centre
Spherical shell	$\frac{2}{3}mr^2$	Through the centre
Uniform disc or cylinder	$\frac{1}{2}mr^2$	About axis of disc or cylinder
Circular band or cylindrical shell	mr^2	About axis of band or cylinder
Uniform rod	$\frac{1}{12}ml^2$	Axis through centre at right angles to rod

Table C2 Selected moments of inertia

You can calculate the moment of inertia of more complicated objects, such as an athlete, by adding the moments of inertia of the individual parts (see Quickfire 13).

quickfire

⑫ For the spheres in Fig. C10, find the moment of inertia for an axis parallel to the one shown which passes through:
a) the centre of mass and
b) the centre of one of the spheres.

quickfire

⑬ Estimate the moment of inertia (about the axis shown) of the upper body of an athlete modelled as in Fig. C11. State your assumptions.

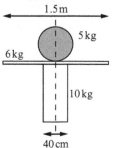

Fig. C11 Composite body

(b) Angular momentum, L

This is defined in the same way as linear momentum – it is equivalent to $p = mv$. In the absence of an external resultant torque (see below), the **angular momentum** of a body is constant – the *principle of conservation of angular momentum*. A rotating athlete, such as a pirouetting ice-skater or a high-board diver, can use this to alter the rate of spin. Going into a tuck reduces the distance of parts of the body from the axis of rotation, hence reducing the moment of inertia and increasing the angular velocity.

Example

The spheres in Fig. C10 are rotating about their centre of mass. An internal mechanism halves their separation. What is the effect on the angular velocity?

Answer

The moment of inertia decreases to $\frac{1}{4}$ of the original because $I \propto r^2$.

∴ Because angular momentum is conserved, ω is multiplied by 4.

(c) Torque, τ

Torque is the rotational equivalent of force. To produce an angular acceleration (without a linear acceleration) we need two equal and opposite forces, offset from each other as in Fig. C12. Such an arrangement is called a *couple*.

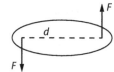

Fig. C12 Couple

Torque is defined as in **Key terms** but the torque of a couple is often conveniently calculated using the equation $\tau = Fd$, which you'll recognise as the moment of the couple about any point.

(d) Kinetic energy

A rotating rigid object has kinetic energy because all the particles in it are moving. The kinetic energy of an individual particle is given, as usual by $\frac{1}{2}mv^2$; the total kinetic energy due to rotation is given by

$$E_{k\,rot} = \tfrac{1}{2}I\omega^2$$

If a body has translational motion as well as rotational motion, e.g. a rolling ball, the total kinetic energy is just the sum:

$$E_k = E_{k\,trs} + E_{k\,rot} = \tfrac{1}{2}mv^2 + I\omega^2$$

Example

A bowling ball of mass of $7.0\,\text{kg}$ is bowled down a lane. When it starts to roll its speed is $6.0\,\text{m s}^{-1}$. Calculate its kinetic energy at that point.

Answer

Assuming the bowling ball is a uniform sphere, $I = \frac{2}{5}mr^2$ $\therefore$

$$E_k = \tfrac{1}{2}mv^2 + \tfrac{1}{2}I\omega^2 = \tfrac{1}{2}mv^2 + \tfrac{1}{2} \times \tfrac{2}{5}mr^2 \left(\frac{v}{r}\right)^2 = \tfrac{7}{10}mv^2$$

$$= 180 \text{ J}$$

C.4 Projectiles

C.4.1 Ignoring lift and drag

We can use the simple projectile theory which we met in Section 1.2.5 of the AS SRG for sports in which a projectile is relatively small and heavy. In fact, this is rather restrictive – probably limited to hammer throwing, cricket, volleyball or one of the variants of boules (see **Pointer**).

Here is a brief recap on the basic theory:

1. Horizontal and vertical components of motion are independent.

2. The horizontal component of velocity is constant.

3. The vertical component of acceleration is downwards and constant (g).

Defining the positive vertical direction as upwards, if the initial horizontal and vertical components of velocity are u_x and u_y, the components at time t later are given by:

$$v_x = u_x \quad \text{and} \quad v_y = u_y - gt$$

Many people take the initial position of the projectile as $(0, 0)$ – even though the projectile is usually thrown from above ground level. The position (x, y) at time t later is given by

$$x = u_x t \quad \text{and} \quad y = u_y t - \tfrac{1}{2}gt^2$$

An example will illustrate these principles.

Example

A cricketer throws a ball at $30\,\text{m s}^{-1}$ at $35°$ from a height of $1.8\,\text{m}$. Where does it hit the ground?

Answer

Consider the vertical motion. Calculate the time of impact:

$u_y = u \sin\theta = 17.2\,\text{m s}^{-1}$; at impact $y = -1.8$ m; $g = -9.81\,\text{m s}^{-2}$

$y - u_y t - \tfrac{1}{2}gt^2$ $\quad \therefore 4.905t^2 - 17.2t - 1.8 = 0$

Grade boost

You don't need to know the radius of a rolling object to calculate its rotational kinetic energy. e.g. for a rolling cylinder, $I = \frac{1}{2}mr^2$.

$\therefore E = \frac{1}{2}mv^2 + \frac{1}{2} \times \frac{1}{2}mr^2\left(\frac{v}{r}\right)^2$

$= \frac{3}{4}mv^2$

quickfire

⑯ The wheels of a bicycle, of total mass $14\,\text{kg}$, each have a mass of $2.8\,\text{kg}$. The cyclist has a mass of $70\,\text{kg}$. Calculate the fraction of the total kinetic energy which is rotational kinetic energy of the wheels.

≫ Pointer

The examiner might well ask you to apply projectile theory to situations in which drag or lift is significant, e.g. javelin, discuss, golf or archery. They might then go on to explore the problems with this approach and possible refinements.

≫ Pointer

If the initial velocity is u at angle θ to the horizontal $u_x = u \cos\theta$ and $u_y = u \sin\theta$.

Grade boost

With an A-level projectiles question, you are likely to have to provide the structure yourself, whereas an AS question will be structured for you.

quickfire

⑰ Calculate the velocity with which the cricket ball in the example hits the ground.

>> *Pointer*

The term 'lift' is not explicitly mentioned in the specification. However, the examiner could ask questions about the physics of lift using Newton's laws.

>> *Pointer*

When considering lift, it is conventional (and easier) to picture the object as stationary in a moving stream of air; the physics is the same. The discus in Fig. C13 is moving to the right in still air but we picture it as the air moving to the left past a stationary discus!

Key Term

The **drag coefficient**, C_D, is a dimensionless constant which relates the drag on an object to its speed and depends upon the shape of the cross section.

Examples:
Sphere   ~0.5
Cube   ~1.0
Flat plate ~1.2
Tear drop ~0.04

Grade boost

Don't take the relative magnitude of the effects of the drag and lift too literally. It all depends upon the design and speed of the projectile.
You should make sure you can discuss the shape of these tracks.

Using the quadratic formula: $t = \dfrac{17.2 \pm \sqrt{17.2^2 + 4 \times 4.905 \times 1.8}}{2 \times 4.905} = 3.61\,\text{s}$

[Ignoring the negative solution]

Consider the horizontal motion. Calculate the distance in time 3.61 s

$\therefore v_x = u_x = u\cos\theta = 24.6\,\text{m s}^{-1}$

Hits the ground at $x = 24.6 \times 3.61 = 89\,\text{m}$ (2 sf)

C.4.2 Lift and drag

These two effects modify the parabolic path of the free projectile in opposite ways: lift tends to extend the path and drag to reduce it.

(a) Lift

Whether in aeroplane flight or sporting contexts, lift arises from the deflection of air as an object moves through it. This is the same mechanism as the force on the rudder in Fig. C5.

As it moves to the right, the discus in Fig. C13 deflects the air downwards. The change in momentum of the air is downwards (and slightly to the right). So (N2) the discus exerts a downwards force on the air and (N3) the air exerts an equal and opposite force on the discus – this force is the lift.

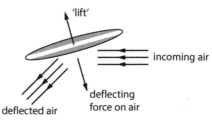
Fig. C13 Lift on a discus

(b) Aerodynamic drag

The movement of an object through the air creates a high-pressure region in front and a low pressure region behind; hence there is a resultant force backwards due to this pressure difference. This is aerodynamic drag. For very low speeds the drag force, F_D, is proportional to the velocity, but in sporting contexts it is proportional to the square of the speed and we have:

$$F_D = \tfrac{1}{2}\rho v^2 A C_D$$

where ρ is the density of the air, A the cross-sectional area of the object and C_D is the **drag coefficient**.

(c) How lift and drag affect projectile motion

Drag slows down motion. Lift is more complicated. Fig. C14 shows the forces (not to scale) on a javelin near the top of its trajectory. Because the javelin is designed so its centre of lift is behind the centre of gravity, the effect of the lift is not only to keep the javelin up but also to rotate it, so that it sticks in

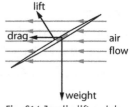
Fig. C14 Javelin lift and drag

the ground when it lands. The combined effect of lift and drag depends on their relative magnitudes. Fig. C15 illustrates their effects.

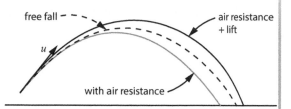

Fig. C15 The effects of drag and lift on projectile motion

C.5 The Bernoulli effect

When a fluid speeds up, its pressure drops. This counter-intuitive result can be derived using the principle of conservation of energy and the relationship between pressure and speed is given in the Data booklet as:

$$p_1 = p_0 - \tfrac{1}{2}\rho v^2 \quad \text{(the Bernoulli equation)}$$

where p_0 is the pressure of the stationary fluid and p_1 its pressure when its speed is v (see **Pointer**).

An example is water being forced through a narrow nozzle in a firefighter's hose (as in Fig. C16). The pressure drop can be great enough to require two firefighters to hold the nozzle steady.

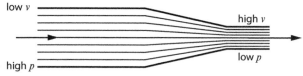

Fig. C16 Fluid flowing through a nozzle

C.5.1 Bernoulli and the Magnus effect

Some people attribute the Magnus effect, which is the veering of a spinning sports ball in the direction of its spin to Bernoulli – see Fig. C17. This is explained in Fig. C18 – the ball is moving to the right but, again, we imagine a stationary ball in a left moving stream of air.

The air is deflected downwards because the air passing over the top, which is travelling *with* the spin, remains in contact with the ball for longer than the underneath air. Compare the two break points A and B.

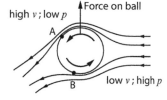

Fig. C18 Magnus effect– Bernoulli explanation

This is the argument:

- The spinning ball diverts the ball downwards ...
- ... so the air travels further over the top than the bottom ...
- ... so the speed is higher at the top than the bottom ...
- ... so the pressure is higher at the bottom than the top.
- The pressure difference causes an upward force on the ball.

>> *Pointer*

You may come across the following alternative ways of writing the Bernoulli equation:

$$p + \tfrac{1}{2}\rho v^2 = \text{constant}$$

$$p_1 + \tfrac{1}{2}\rho v_1{}^2 = p_2 + \tfrac{1}{2}\rho v_2{}^2$$

If the vertical level of the fluid changes, there is an additional term of ρgh in the equation.

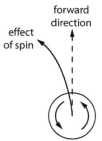

Fig. C17 Magnus effect

>> *Pointer*

The main problem with the Bernoulli explanation is that it overestimates the speed of air on the side with the spin. Measurements suggest that it is larger than the underneath speed but only by a small fraction.

quicKfire

⑱ Explain the direction of the Magnus effect using N2 and N3.

C.5.2 Newton's laws and the Magnus effect

The Bernoulli 'explanation' for the Magnus effect is now regarded as only accounting for a small fraction of the observed force. Newton's own explanation (supposedly given on observing a tennis match in Trinity College, Cambridge) was just the same as the working of the rudder in Section C.2.1 – it still relies on the diversion of the air and for the same reasons.

1. An Olympic diver changes her form from a 2 m straight line to a 1 m tucked shape, which we'll model as a uniform straight rod and a uniform disc respectively.

   (a) Explain why her rotation speeds up when she does this.

   (b) Calculate by what factor her rotation speeds up.

2. A cyclist of mass 60 kg rides a bicycle of mass 18 kg. The two road wheels each have a mass of 2.5 kg. The cyclist free-wheels down a 15 m high slope of length 200 m from an initial velocity of $5.0\,\text{m}\,\text{s}^{-1}$.

   (a) Ignoring frictional losses, calculate the speed of the cyclist at the bottom of the slope.

   (b) Compare this speed with that achieved by a friction-free sledge sliding down a similar slope.

   (c) Calculate the acceleration of the cyclist.

   (d) The road wheels have a radius of 66.0 cm. Calculate the angular acceleration of the wheels.

3. A model sports drone has 4 horizontal rotors, of radius 10 cm as shown. The mass of the rim is 8 g and the total mass of the rotor blades is 12 g. On takeoff it is spun up to 5000 rpm in 5 seconds. Calculate:

   (a) the moment of inertia of each rotor

   (b) the mean resultant torque on the rotors during spin-up

   (c) the total kinetic energy of the 4 rotors.

   [For a rod rotated about its end, $I = \frac{1}{3}ml^2$]

20 cm

4. The drone in question 3 has a mass of 1 kg. It can lift off because each rotor pushes a 10 cm radius column of air downwards. Calculate:

   (a) the speed of the air needed to support the drone

   (b) the power required.

   [Density of air $= 1.3\,\text{kg}\,\text{m}^{-3}$]

5. Pairs of the rotors in question 3 are arranged to rotate in opposite directions. Suggest why this is an advantage.

6. A shot is put (thrown) at 42° to the horizontal and $12.0\,\text{m}\,\text{s}^{-1}$ from an initial height of 1.90 m.

   (a) Calculate the range of the shot.

   (b) His trainer suggests he switch to an angle of 36° at which he can achieve a putting speed of $13.0\,\text{m}\,\text{s}^{-1}$. Will this improve his result? If so, by how much?

   (c) Briefly explain the result qualitatively.

Option D: Energy and the environment

This optional area of study contains two major themes:

- An introduction to the physics of atmospheric warming
- Sources of energy

and minor themes of fuel cells and thermal conduction. Each has its own subsection.

D.1 The Earth's rising temperature

The temperature of the Earth's atmosphere and surface (land and sea) is rising. It is accepted that this warming is largely anthropogenic in origin. This section explores some of the influences on global temperatures.

D.1.1 The need for thermal equilibrium

The main energy input to the Earth's surface is solar radiation. Measured at the top of the atmosphere, the intensity of the radiation is $1.37\,\mathrm{kW\,m^{-2}}$, from which we can calculate the total solar energy input. Fig. D1 shows how the visible part of this radiation interacts with the Earth and atmosphere – about 30% is reflected and the rest absorbed.

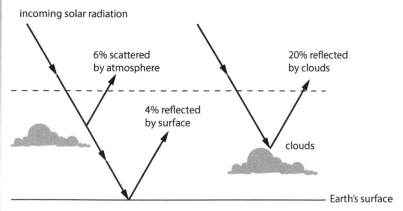

incoming solar radiation

6% scattered by atmosphere

20% reflected by clouds

4% reflected by surface

clouds

Earth's surface

Fig. D1 Reflection of solar radiation by the Earth and atmosphere

》 Pointer

The intensity of the radiation from the Sun is not strictly a constant. The intrinsic variation of solar radiation is about ~0.1%. It also changes because the Earth–Sun distance varies.

quickΦιρε

① Show that the total solar energy input to the Earth per second is approximately $2 \times 10^{17}\,\mathrm{W}$.

$[R_\mathrm{E} = 6370\,\mathrm{km}]$

quickΦιρε

② Heat is conducted from the centre of the Earth at an estimated rate of 44 TW. What fraction of the solar energy input does this represent?

Fig. D1 cannot be the whole story or the earth and atmosphere would be rising in temperature the whole time. To keep at a constant temperature, the Earth must emit energy so that the net energy input is zero. It emits infra-red radiation (see Section D.1.2) which is partly absorbed by the atmosphere and re-radiated in all directions.

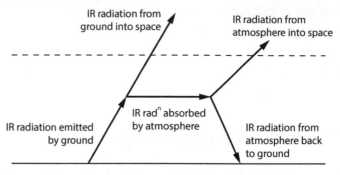

Fig. D2 IR radiation emission and absorption

IR radiation is partly absorbed by polyatomic molecules, especially water vapour, carbon dioxide (CO_2) and methane (CH_4). The result of the re-emission of radiation by the atmosphere and its subsequent absorption by the ground raises the temperature of Earth to a higher level than it otherwise would be. This is called the greenhouse effect and is a natural process. CO_2 and CH_4 are present at higher levels as a result of human activity, leading to greater absorption of IR, an enhanced greenhouse effect and thus resulting in global warming.

D.1.2 The spectra of radiation from the Sun and Earth

The Sun's photosphere (its visible disc) has a temperature of 5770 K. Using Wien's law (see the AS SRG, Section 1.6.3) we can work out the peak wavelength of its spectrum. The solar spectrum is shown in Fig. D3.

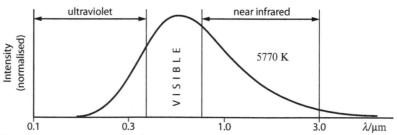

Fig. D3 The solar spectrum

The atmosphere is transparent to the near UV ($\lambda > 0.3\,\mu\text{m}$), visible and near IR ($\lambda < 2\,\mu\text{m}$) which accounts for most of the incoming and reflected radiation. By comparison the Earth's surface has a mean temperature of 288 K, so its radiation is in the range $3 - 100\,\mu\text{m}$, and is quite strongly absorbed and re-radiated by H_2O, CO_2, CH_4, and N_2O molecules.

Example

Explain briefly why increasing CO_2 levels in the atmosphere leads to global warming.

Answer

The ground emits long wavelength IR radiation which is partly absorbed by CO_2 molecules in the atmosphere. These molecules then re-emit the radiation, some downwards which is absorbed by the ground and raises its temperature.

The more CO_2 there is in the atmosphere, the more this radiation is absorbed and re-emitted, thus increasing the energy received by the ground and raising its temperature further.

D.1.3 Temperature of a planet without a greenhouse effect

If a planet absorbs a fraction μ of the radiation incident upon it we can estimate its temperature using Stefan's law (see AS SRG Section 1.6.4) using the value of the solar constant, I. The principle is that, for equilibrium the net power input from the Sun must be balanced by an equal power output from the planet.

The area presented by the planet to the Sun's radiation is πR^2, where R is the radius of planet. The planet emits radiation from its entire surface, i.e. from an area $4\pi R^2$. If the planet's temperature is T:

Power absorbed $= \mu\pi R^2 I$ and   Power emitted $= 4\pi R^2 \sigma T^4$

$\therefore$ For equilibrium $4\pi R^2 \sigma T^4 = \mu\pi R^2 I$

So, simplifying and rearranging $T^4 = \dfrac{\mu I}{4\sigma}$

Example

Use the above equation to estimate the temperature of the Earth without an atmosphere.

Answer

The Earth's reflects 30% of the light (see Fig. D1), so $\mu = 0.7$

$$\therefore T^4 = \frac{\mu I}{4\sigma} = \frac{0.7 \times 1370 \ \text{W m}^{-2}}{4 \times 5.67 \times 10^{-8} \ \text{W m}^{-2}\text{K}^{-4}} = 4.2 \times 10^9 \ \text{K}^4$$

$\therefore T = 255\,\text{K}$

Discussion

The mean surface temperature of the Earth is actually $288\,\text{K}$, so the greenhouse effect produces a roughly $30\,\text{K}$ (i.e. $30\,°\text{C}$) gain in temperature. The greenhouse gases in the atmosphere raise the surface temperature to above $0\,°\text{C}$, allowing life to exist in the form we know it.

Grade boost

Don't bother to learn the equation which we have derived; just understand the ideas.

quickfire

⑥ Explain why the working here assumes that the planet is rotating quite quickly.

Pointer

Note that the working of the example takes scattering from the air into account but not the greenhouse effect. It assumes an atmosphere without CO_2, H_2O, CH_4, etc. But see Quickfire 7

quickfire

⑦ If there is no water in the atmosphere there can't be any clouds. Find an estimate for the Earth's temperature ignoring the reflection from the clouds (see Fig. D1). You should find it is still below $0\,°\text{C}$.

D.1.4 Global warming and rising sea levels

Currently, sea levels are rising at about 3 mm per year due to global warming. There are two reasons for this:

1. Thermal expansion of the sea.

2. Melting land ice (but not sea ice).

Archimedes' principle explains why the melting of floating sea ice has no effect on sea levels.

(a) Archimedes' principle

The floating ice in Fig. D4 *displaces* a volume of seawater – this is the volume of the iceberg which is submerged.

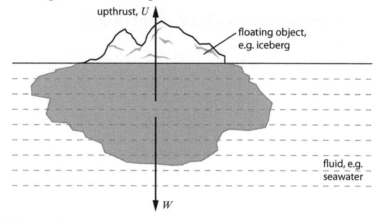

Fig. D4 Forces on floating ice

If the volume of the berg is V, what volume is submerged?

Weight of berg, $W = \rho_i V g$ where ρ_i = density of ice

Using Archimedes' principle, If the submerged volume is V' the upthrust, U, is given by:

$$U = \rho_w V' g \qquad \text{where } \rho_w = \text{density of seawater}$$

If the berg is floating at equilibrium, $U = W$ so $\rho_w V' g = \rho_i V g$

Hence, rearranging and simplifying: $V' = \dfrac{\rho_i V}{\rho_w}$

So, because $\rho_i < \rho_w$ (see Quickfire 8) an iceberg floats with some of its volume above the water surface.

(b) Why melting icebergs don't raise the sea level

Where does the water go that was above the sea level in the floating berg? The iceberg becomes liquid water, with the same density as the seawater. Hence it 'floats' with 100% of its volume submerged, i.e. the melting iceberg shrinks to occupy just the submerged volume! But melting sea ice does have positive feedback effect on global temperatures (see Quickfire 9).

Key Term

Archimedes' principle states that a body which is wholly or partially immersed in a fluid experiences an upthrust equal to the weight of the fluid displaced by the body.

quickfire

⑧ The density of ice is $917\,\mathrm{kg\,m^{-3}}$; the figure for North Atlantic sea water is $1030\,\mathrm{kg\,m^{-3}}$. It is often stated that icebergs have 90% of their volume submerged. Is this correct?

quickfire

⑨ Floating ice is white; when it melts, it leaves the 'wine-dark sea' (Homer's *Odyssey*). Explain why the melting of the Arctic icecap further raises the mean global temperature.

D.2 Energy sources

The specification mentions renewable and non-renewable sources but all the detail is about the former. Hence you should be able to use your GCSE knowledge of non-renewable sources (coal, oil, natural gas) especially in the light of their carbon dioxide emissions and global warming but the additional knowledge is confined to the named sources in the specification.

D.2.1 Solar power

(a) Nuclear fusion in the Sun

Approximately 98% of the energy output of the Sun comes from the **proton–proton chain**. It dominates for stars of less than 1.3 solar masses. The energy produced comes from the binding energy (see Section 3.6) released in a set of reactions which can be summarised as:

$$4{}^1_1H + 2{}^{\ 0}_{-1}e \rightarrow {}^4_2He + 2\nu_e \qquad \text{pp chain summary}$$

with *nuclear* symbols, i.e ${}^1_1H$ = proton, ${}^4_2He$ = helium nucleus

There are several routes but they all start with

$$ {}^1_1H + {}^1_1H \rightarrow {}^2_1H + {}^0_1e+ + \nu_e \qquad \text{pp stage 1}$$

followed by

$$ {}^1_1H + {}^2_1H \rightarrow {}^3_2H + \gamma \qquad \text{pp stage 2}$$

where ${}^2_1H$ is a heavy hydrogen (deuterium) nucleus (see **Pointer**).

Stage 1 is a very slow reaction but stage 2 is much quicker. You should be able to explain why – Hint: strong/weak.

After stage 2 there are several routes to ${}^4_2He$. The main one, known as ppl is

$$ {}^3_2H + {}^3_2H \rightarrow {}^4_2He + {}^1_1H + {}^1_1H \qquad \text{ppl stage 3}$$

Where do the electrons in the summary equation come into things? To get to the two **He-3** nuclei, two stage 1 reactions need to occur, each of which produces a positron, e^+. Each positron undergoes annihilation with an electron (of which there are many) in the Sun's core.

(b) The solar constant

The total power, P, released by the Sun is 3.846×10^{26} W. From this you can calculate the number of fusion reactions occurring every second as well as the mass loss per second using $E = mc^2$. The radiation carrying this energy spreads out spherically symmetrically, so that, at any distance, r, the intensity of the solar radiation is given by

$$I = \frac{P}{4\pi r^2}$$

Hence the intensity falls off according to the inverse square law.

» Pointer

Solar power is indirectly the source of most renewable energy sources, including wind, waves, hydro and biomass.

Key Term

The **proton-proton chain** is a set of reactions starting with the fusion of two protons to deuterium.

» Pointer

${}^2_1H$ can also be written ${}^2_1D$. Similarly ${}^3_1H$ can be written ${}^3_1T$ (T = tritium)

quickfire

⑩ Calculate the energy released when two protons undergo the pp stage 1 reaction. Express your answer in:
a) MeV
b) J
Mass data:
${}^1_1H$ = 1.007 276 u
e^+ = 5.49×10^{-4}u
${}^2_1H$ = 2.013 562 u

quickfire

⑪ Use your answer to Quickfire 10 to calculate the energy released (in J) by a mole of protons.

(Care: There's a trap here!)

» Pointer

The inverse square law for radiation is equally applicable to point sources and to spherically symmetric sources.

quicĸꭲɪꭱe

⑫ The radius of the Sun is 696,000 km. Use Stefan's law to calculate the temperature of the photosphere. Compare this with the data already given.

≫ Pointer

$I\cos\theta$ is the component of the radiation flux at right angles to the PV panel. In Britain, many roofs are at 35–40° to the horizontal and the maximum height of the Sun at midday is about 60° in summer and 15° in winter.

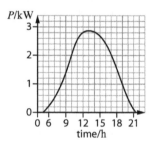

Fig. D6 PV panel output

quicĸꭲɪꭱe

⑬ Estimate the total day's output of the solar panel shown in Fig. D6.

Example

The mean radius of the Earth's orbit is 149.6 million km. Calculate the **solar constant**.

Answer

Intensity at the Earth's distance $= \dfrac{3.846 \times 10^{26}\,\text{W}}{4\pi \times (1.496 \times 10^{11}\,\text{m})^2}$

i.e.   Solar constant $= 1367\ \text{W m}^{-2}$

(in agreement with previous information!)

(c) Photovoltaic cells

The conversion efficiency of a PV panel is of the order of 25%. If the efficiency is μ, the power output, P, of a solar panel of area A, in sunlight of intensity I, is given by

$$P = \mu A I \cos\theta$$

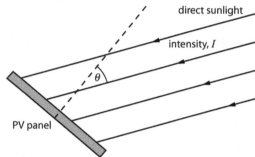

Fig. D5 Solar panel input

where θ is the angle between the normal and the direction of the incident sunlight.

Example

On a summer day in Britain, the intensity of the sunlight is $650\ \text{W m}^{-2}$ and the Sun's elevation is 60°. Calculate the power output of a $20\ \text{m}^2$ array of solar panels at 35° to the horizontal if the conversion efficiency is 22%.

Answer

Angle between the sunlight and the solar panel normal = 5°

∴ Electrical power output $= 0.22 \times 20\ \text{m}^2 \times 650\ \text{W m}^{-2}\cos 5°$

$= 2800\ \text{W (2 sf)}$

D.2.2 Wind power

Consider the cylinder of air moving towards the wind turbine in Fig. D7. The kinetic energy in the air which reaches the turbine in Δt is given by:

$$E_\text{K} = \tfrac{1}{2}mv^2 = \tfrac{1}{2}(\rho A v \Delta t)v^2 = \tfrac{1}{2}\rho A v^3 \Delta t$$

where A is the cross-sectional area of the cylinder.

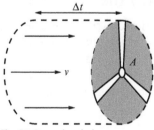

Fig. D7 Power in wind

So the power available, $P = \dfrac{E_\text{K}}{\Delta t} = \tfrac{1}{2}\rho A v^3$

The maximum power output of the turbine is lower than this for the following reasons:

- There is a lower 'cut-in' wind speed for the turbine to operate.
- The turbine blades turn to restrict the power above the rated wind speed.
- The turbines cut out at even higher wind speeds (see fig. D8).
- Wind speed varies rapidly – the turbine cannot respond instantly.

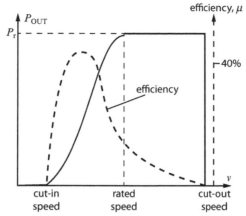

Fig. D8 Typical characteristic of a wind turbine

- The air doesn't lose all its kinetic energy when it passes through the turbine.
- The turbines in a wind farm interfere with one another.

Tidal stream turbines

These schemes work on the same principle as wind farms. The underwater turbines are placed in regions with a high-speed tidal flow. The energy content of moving water is higher than air because of the higher density. Another advantage is the predictability of the tides – with four changes of tide per day.

D.2.3 Electricity from the potential energy of water

Hydroelectric power stations, including pumped storage schemes and tidal barrages operate by capturing the loss in potential energy of water as it flows downhill (Fig. D9).

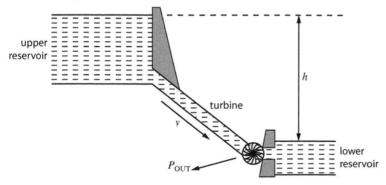

Fig. D9 Structure of a hydroelectric system

Without a turbine, the energy change in the system is just the familiar gravitational potential to kinetic of a falling object. So we can use $\frac{1}{2}mv^2 = mgh$ giving $v = \sqrt{2gh}$ to calculate the volume flow rate (Quickfire 14) and hence the energy transfer.

quickfire

⑭ Water flows down a 50 cm diameter pipe from an upper to a lower reservoir, with a height difference of 10 m.

a) Show that the speed of the water is about $14\,\mathrm{m\,s^{-1}}$.

b) Calculate the volume flow rate.

quickfire

⑮ Show that you get the same answer for energy transfer rate in Example 1 using $\frac{1}{2}mv^2$.

Example 1

Calculate the rate of energy transfer in Quickfire 14.

Answer

The answer to Quickfire 14 (b) is $2.75\,\text{m}^3\,\text{s}^{-1}$.

So the mass transfer rate $= 2.75\,\text{m}^3\,\text{s}^{-1} \times 1000\,\text{kg}\,\text{m}^{-3} = 2750\,\text{kg}\,\text{s}^{-1}$

$\therefore$ Using loss of PE $= mgh$, energy transfer rate $= 2750 \times 9.81 \times 10\,\text{W}$

$$= 270\,\text{kW}$$

In a practical power station, a turbine is built into the pipe as in Fig. D9. This restricts the flow rate so the water does not gain as much kinetic energy because it has to do work on the turbine. Hence:

$$\text{PE loss} = \text{KE gain} + \text{work done on turbine}$$

Example 2

When a turbine is built into the set-up in Quickfire 14, the water speed is reduced to $10\,\text{m}\,\text{s}^{-1}$. Assuming no losses, calculate the power output of the turbine.

Answer

The mass flow rate $= \pi r^2 v \rho = 1960\,\text{kg}\,\text{s}^{-1}$

$\therefore$ Rate of PE loss $= mgh = 1960 \times 9.81 \times 10 = 192\,\text{kW}$

and rate of KE gain $= \frac{1}{2}mv^2 = \frac{1}{2} \times 1960 \times 10^2\,\text{W} = 98\,\text{kW}$

$\therefore$ Power output of turbine $= 192\,\text{kW} - 98\,\text{kW} = 94\,\text{kW}$

quickfire

⑯ If the efficiency of the generator that the turbine is connected to is 95%, calculate the efficiency of the power station in Example 2.

D.2.4 Nuclear fission and fusion

(a) Nuclear fuel enrichment

Natural uranium consists of 99.3% non-fissile U238 and 0.7% U235. The U235 content needs to be increased to between 3% and 5% for most civil reactors to operate. **Enrichment** is achieved by reacting natural uranium with fluorine to produce the gas uranium hexafluoride (UF_6) and feeding this into a cascade of gas centrifuges. Partial separation is achieved because of the difference in molecular mass of the two isotopes – the UF_6 with the U238 atom is spun preferentially to the outside allowing the gas enriched with U235 to be drawn off from the middle. This is converted back to metallic uranium or uranium oxide for use in the reactor.

> **» Pointer**
>
> The details of the construction of fission and fusion reactors are not part of the course. GCSE knowledge is sufficient. A few technical points are picked out.

(b) Nuclear fuel breeding

The non-fissile U238 in the fuel rods is transformed into the fissile **Pu239** by the capture of a neutron followed by two β^- decays:

$$^{238}_{92}\text{U} + ^{1}_{0}\text{n} \rightarrow ^{239}_{92}\text{U} \xrightarrow[23.5\ \text{min}]{\beta^-} ^{239}_{93}\text{Np} \xrightarrow[2.3\ \text{day}]{\beta^-} ^{239}_{94}\text{Pu}$$

> **» Pointer**
>
> A significant fraction of the power produced in a reactor comes from the fission of Pu239 – the fraction increases during the fuel element's life.

At the reprocessing stage, **Pu239** is extracted and can be incorporated in new fuel elements. Pressurised water reactors operate with up to 30% plutonium in the mixed oxide fuel (MOX).

(c) Nuclear fusion triple product

In order to obtain a sustained fusion reaction, three conditions need to be fulfilled at the same time:

1. A high enough temperature, T, so that the colliding nuclei can approach close enough against the Coulomb repulsion to allow fusion to occur.

2. A high particle density, n, so that the number of collisions between nuclei is great enough.

3. A long confinement time, τ_E (see Quickfire 18).

Achieving these three conditions together is very difficult. Temperatures of the order of $100\,\text{MK}$ are needed but the plasma must be kept away from the container walls, so that the energy is not lost. The **fusion triple product** is the product of these three quantities, $nT\tau_E$. From theoretical studies, engineers know what value the triple product must exceed for any reaction to occur. For example, the deuterium–tritium reaction is thought of as a likely practical reaction for a fusion reactor:

$$^2_1\text{H} + {}^3_1\text{H} \rightarrow {}^4_2\text{He} + {}^1_0\text{n}$$

The triple product required is approximately $3.5 \times 10^{28}\,\text{K}\,\text{m}^{-3}\,\text{s}$.

quicKfire

⑰ The confinement time, τ_E, is defined by
$$\tau_E = \frac{W}{P_{\text{loss}}}$$
where W is the energy density and P_{loss} is the power loss per unit volume. Show that τ_E has units of time.

quicKfire

⑱ A prototype fusion reactor can obtain values of $n\tau_E$ as high as $3 \times 10^{20}\,\text{m}^{-3}\,\text{s}$. Calculate the temperature required.

D.3 Fuel cells

A fuel cell works like a battery which is always being topped up with its fuel chemicals. For vehicles, most use hydrogen and oxygen (from the air). Very schematically:

1. Hydrogen is drawn in at the anode and ionised (i.e. electrons removed) by a platinum catalyst. The protons diffuse across an insulating barrier.

2. The electrons travel around an external circuit where they do work and then re-enter the fuel cell at the cathode where they combine with oxygen and the diffused protons to produce water. This reaction also requires a catalyst.

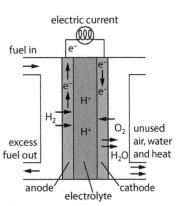

Fig. D10 Fuel cell

Advantages of the fuel cell

- High efficiency – theoretically up to 80%; 40% is regularly achieved.
- Only water vapour produced – no CO_2.
- Drives electric cars so simplifying the gearing required.

Note that the second advantage is only real if the hydrogen is produced without causing greenhouse emissions, e.g. by electrolysis using electricity from PV cells.

» Pointer

Note the minus sign in the conduction equation. Heat flows *down* a temperature gradient, i.e. from a high to a low temperature.

D.4 Thermal conduction

D.4.1 The conduction equation

The rate of transfer of energy by heat conduction, $\frac{\Delta Q}{\Delta t}$, through a sample of material, depends upon the temperature difference, $\Delta\theta = \theta_2 - \theta_1$, the thickness, x, the cross-sectional area, A, and the nature of the material.

The relationship is $\frac{\Delta Q}{\Delta t} = -AK\frac{\Delta\theta}{\Delta x}$

where K is a constant called the coefficient of thermal conductivity of the material.

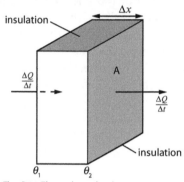
Fig. D11 Thermal conduction

It is worth noting that the orders of magnitude of K for different kinds of material which are used in engineering:

Material	Coefficient of thermal conductivity order of magnitude / $W\ m^{-1}\ K^{-1}$
Gases (air / argon)	10^{-2}
Concrete / brick / ceramic	1
Metals	10^2

Table D1 Coefficients of thermal conductivity

The only tricky problems involving this equation have two materials in contact.

» Pointer

The principle of solving the example is that, because of the insulation, the rate of heat transfer is the same in the copper and aluminium sections of the bar.

Example

Calculate the rate of heat conduction along a $4\,cm^2$ c.s.a. insulated bar of metal consisting of 5 cm lengths of copper and aluminium 'in series', the ends of which are maintained at 40 °C and 10 °C (see Fig. D12).

$[K_{Cu} = 385\ W\,m^{-1}\,K^{-1} ; K_{Al} = 200\ W\,m^{-1}\,K^{-1}]$

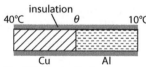
Fig. D12 Composite bar

Answer

If the temperature of the junction is θ,

Rate of heat conduction in the Cu = rate of heat conduction in the Al, so

$$4\,\text{cm}^2 \times 385\,\text{W}\,\text{m}^{-1}\,°\text{C}^{-1} \times \frac{40\,°\text{C} - \theta}{5\,\text{cm}} = 4\,\text{cm}^2 \times 200\,\text{W}\,\text{m}^{-1}\,°\text{C}^{-1} \times \frac{\theta - 10\,°\text{C}}{5\,\text{cm}}$$

$\therefore$ (dispensing with the units) $385(40 - \theta) = 200(\theta - 10)$

Which leads to $\theta = 29.7\,°\text{C}$

Substituting into $\dfrac{\Delta Q}{\Delta t} = -AK\dfrac{\Delta \theta}{\Delta x}$ for the copper part of the bar:

$$\frac{\Delta Q}{\Delta t} = -4 \times 10^{-4}\,\text{m}^2 \times 385\,\text{W}\,\text{m}^{-1}\,°\text{C}^{-1}\frac{(29.7 - 40)\,°\text{C}}{0.05\,\text{m}} = 31.7\ \text{W}$$

» Pointer
Because $\Delta\theta$ is a temperature difference, the unit of K can be either $\text{W}\,\text{m}^{-1}\,\text{K}^{-1}$ or $\text{W}\,\text{m}^{-1}\,°\text{C}^{-1}$.

quickfire
⑲ Show that we get the same answer for the rate of heat flow using the aluminium part of the bar.

D.4.2 *U* values

The *U* value is a rating for heat loss through unit area of a construction structure (rather than a material) per unit temperature difference under standard conditions (see **Pointer**).

$$\frac{\Delta Q}{\Delta t} = -UA\Delta\theta$$

The *U* value cannot easily be related to the coefficient of thermal conductivity, even for a simple structure, such as a solid wooden door. This is because part of the thermal insulation is due to a stationary layer of air in contact the inside and outside surface. Usually a wall contains other structures, typically windows and doors, which act like components in parallel: the total heat loss is the sum of the heat loss through each structure.

» Pointer
U values are usually quoted for a 24 °C temperature difference and 50% humidity.

Example

Calculate the rate of heat loss through the insulated cavity wall, window and door, shown in Fig. D13. The door and double glazed window each have an area of $2.0\,\text{m}^2$; the internal and external temperatures are 20 °C and 12 °C respectively.

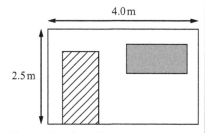

Fig. D13 Exterior wall

quickfire
⑳ What would be the percentage saving on the heating bills if a new insulated door with a *U* value of $1.4\,\text{W}\,\text{m}^{-2}\,\text{K}^{-1}$ were installed?

$$U_{\text{wall}} = 0.24\,\text{W}\,\text{m}^{-2}\,\text{K}^{-1};\ U_{\text{window}} = 1.2\,\text{W}\,\text{m}^{-2}\,\text{K}^{-1};\ U_{\text{door}} = 3.0\,\text{W}\,\text{m}^{-2}\,\text{K}^{-1}$$

Answer

Area of wall = $4.0\,\text{m} \times 2.5\,\text{m} - 2.0\,\text{m}^2 - 2.0\,\text{m}^2 = 6.0\,\text{m}^2$

$$\therefore \frac{\Delta Q}{\Delta t} = (0.24 \times 6.0 + 1.2 \times 2.0 + 3.0 \times 2.0) \times (20 - 12)\,\text{W} = 79\ \text{W}$$

1. Use the data below to estimate the surface temperature of Mars. Assume that the thin Martian atmosphere has a negligible greenhouse effect.
   - Power output of the Sun $= 3.846 \times 10^{26}$ W.
   - Mean orbital distance of Mars $= 227.9$ million km.
   - Mars reflects 29% of the Sun's radiation.

2. Use your answer to question 1 to estimate the peak wavelength of the radiation emitted by Mars.

3. A wind turbine has blades of length 35.4 m. It is in a steady wind of speed $8.0\,\mathrm{m\,s^{-1}}$. Calculate:
   (a) the kinetic energy density in the wind (i.e. the kinetic energy per unit volume)
   (b) the volume of air incident upon the wind turbine per unit time
   (c) the power output of the turbine given that its efficiency is 40%
   (d) the blade length required for a tidal stream turbine to produce the same power output in a water stream of the same speed. (Assume the same efficiency.)
   [Density of air $= 1.23\,\mathrm{kg\,m^{-3}}$; density of sea water $= 1025\,\mathrm{kg\,m^{-3}}$]

4. As the Siberian tundra (frozen boggy ground) thaws out, it decomposes and releases methane. How is the effect of this on the atmosphere similar to that of the melting Arctic ice cap?

5. In a hydro-electric scheme water drains a vertical distance of 120 m down a 1 m diameter pipe from an upper reservoir to a lower reservoir.
   (a) Calculate the speed of flow of the water and hence the rate of energy loss from the upper lake if there is no turbine in the pipe.
   (b) Explain briefly why the maximum power output of a turbine built into the pipe is much less than that calculated in part (a).
   (c) A turbine is built into the pipe which restricts the speed of flow to $20\,\mathrm{m\,s^{-1}}$. Calculate:
      (i) the electrical power output if the turbine/generator combination has an efficiency of 90%
      (ii) the overall energy efficiency of the operation of the scheme.

6. A PV panel of area $10\,\mathrm{m^2}$ is placed horizontally at the equator. On a March day, the Sun rises due East, and moves through the zenith and sets in the West. The intensity of the sunlight is $600\,\mathrm{W\,m^{-2}}$.
   (a) Sketch a graph of the power output of the panel over the course of the day, if the conversion efficiency is 25%.
   (b) Use your graph to estimate the total energy generated by the PV panel over the day.
   (c) Suggest why your graph is likely to over-estimate the power available in the early morning and late afternoon.

7. A hollow steel sphere of radius 50 cm, with walls of thickness 1 cm, contains water at 50 °C. It is placed in the sea, which has a temperature of 10 °C.
   (a) Calculate the initial rate of loss of heat and the initial rate of decrease of temperature. [You may ignore the heat capacity of the steel.]
   (b) State what the two rates will be when the temperature of the water in the sphere has fallen to 30 °C. Explain your answer.
   (c) Without further calculation, sketch a graph of the variation of the water temperature with time and explain its shape.
   [$c_\mathrm{W} = 4200\,\mathrm{J\,kg^{-1}\,K^{-1}}$; $K_\mathrm{steel} = 50\,\mathrm{W\,m^{-1}\,K^{-1}}$]

Component 3 Options Summary

A: Alternating currents

- Quantitative application of Faraday's law to a rotating coil in a magnetic field; $N\Phi = BAN \sin \omega t$ if the angle between the field and coil normal is given by $\theta = \omega t$; hence $V = -BAN \cos \omega t$

- The frequency, period, peak values and rms values of alternating potential differences and currents; for sinusoidal AC $I_{rms} = \dfrac{I_0}{\sqrt{2}}$ and $V_{rms} = \dfrac{V_0}{\sqrt{2}}$

- The use of rms values in power dissipation calculations

- The use of an oscilloscope to measure the frequencies and values of AC and DC currents and voltages

- Phase relationship between pd and current for resistors, capacitors and inductors; capacitors and inductors dissipate no power

- The terms resistance, reactance and impedance applied to components and combinations of components

- $X_L = \omega L$; $X_C = \dfrac{1}{\omega C}$; $$Z_{RCL} = \sqrt{R^2 + \left(\omega L - \dfrac{1}{\omega C}\right)^2}$$

- Using phasors to analyse RCL series circuits, including calculating the phase angle between the current and supply potential difference

- The series RCL resonance circuit; the derivation of the resonance frequency, $f_0 = \dfrac{1}{2\pi\sqrt{LC}}$

- The Q factor defined by $Q = \dfrac{V_L}{V_R} = \dfrac{V_C}{V_R}$ at resonance; Q determines the sharpness of a resonance circuit

B: Medical physics

- The nature, properties and production of X-rays

- The use of X-rays in diagnosis and treatment

- Attenuation of X-rays, $I = I_0 e^{-\mu x}$

- X-ray imaging and fluoroscopy; radiography techniques including digital imaging; CT scans

- Generation and detection of ultrasound using piezoelectric transducers; acoustic impedance, the need for a coupling medium

- Ultrasound diagnosis: A-scans and B-scans; examples and applications

- Doppler scans for studying blood flow

- The principles of magnetic resonance: precession of nuclei, resonance, relaxation time, the Larmor frequency $f = 4.26 \times 10^6 B$

- The use of MRI in diagnosis

- Comparison of ultrasound, X-ray and magnetic resonance imaging for the investigation of internal structures

- The effects of α, β and γ on living matter

- Measures and units of dose including absorbed dose, D, equivalent dose, H and effective dose, E, radiation weighting factor, W_R, tissue weighting factor, W_T, the gray (Gy) and sievert (Sv); the relationships $H = DW_R$ and $E = HW_T$

- Radionuclide tracers for body imaging; Tc99m

- Gamma camera: collimator, scintillation counter, photomultiplier / CCD

- PET scanning; tumour detection

C: The physics of sports

- Use of the principle of moments in the context of stability and toppling, the skeleton-muscular system and sporting contexts
- The application of Newton's 2nd law of motion and the coefficient of restitution
- Moment of inertia; angular momentum and rotational kinetic energy
- Rotational kinematics: angular velocity and acceleration
- Rotational dynamics; torque, conservation of angular momentum
- Conservation of energy in sporting contexts
- Projectile motion in sporting contexts
- Bernoulli's equation
- Drag force and drag coefficient

D: Energy and the environment

- Factors affecting the (rise of) temperature of the Earth
  - Thermal equilibrium
  - Solar energy interactions with the earth and atmosphere
  - The laws of Wien and Stefan Boltzmann
  - Floating ice and land ice
- Renewable and non-renewable energy sources
  - The origin and intensity of solar power; photovoltaic cells
  - Wind power; wind turbines
  - Tidal barrages, hydroelectricity and pumped storage
  - Nuclear fission and fusion; enrichment; fusion triple product
- Fuel cells: principles and benefits
- Thermal conduction; coefficient of thermal conductivity; U values

P Practical work

Practical work is an integral part of the Eduqas A level physics course. Candidates are required to undertake and keep records of a set of practical activities covering part (b) of Appendix A of the specification in the context of the practical techniques specified in part (c). This is known as the non-exam assessment (NEA). In order to undertake this course you will need to demonstrate data-handling skills: Your attention is drawn to the following additional sources of information:

- Section 3 of the AS Study and Revision Guide
- The Maths and Data section of this guide
- The data analysis questions in the end of section exercises in both the AS and A2 Student books

Components 1, 2 and 3 of the examination also include questions on practical work, including recall of techniques, data analysis and evaluation.

P.1 Skills to be developed

The non-exam assessment (NEA) and the component tests both require you to analyse, evaluate and draw conclusions from data. The major difference between the two is that, whilst in one (the NEA) you will obtain your own data, in the other (component tests) it will be given to you.

P.1.1 Graph plotting

You are expected to decide how to plot data to test and given or suggested relationship. In almost all cases this will mean trying to obtain a straight-line plot from the data. This is called *linearising* the data. The techniques for this are covered in Section 3.4 of the AS Study and Revision Guide and in Section M.3 of this book. For some activities this will involve the use of log plots.

P.1.2 Uncertainty analysis

You will need to

- use repeated readings or the resolution of an instrument to estimate the uncertainty in data
- plot error bars using uncertainty in data (this is not required in log graphs)

>> *Pointer*

Candidates for the WJEC A level physics qualification are required to take a practical examination consisting of an Experimental Task and a Data Analysis Task. For details see the specification and the WJEC A2 Physics SRG.

quickfire

① Give suitable graphs to plot to linearise the data for the following relationships, where the variables are x and y:

a) $y = ax^2 + b$

b) $y^2 = kx^n$

c) $y = Ae^{-kx}$

quickfire

② The length, width, and thickness of a metal block are determined to be 10.5 cm, 4.6 cm and 2.2 cm respectively. If the uncertainty in each value is ± 0.1 cm, calculate the volume of the block together with its absolute uncertainty.

quickfire

③ For the relationship in Quickfire 1 (a), state how you would use your graph to find the values of the constants, a and b.

quickfire

④ A graph of ln y against ln x is a straight line with gradient 2.5 and intercept 1.8 on the ln y axis. Determine the relationship between x and y.

(See Section M.3)

- use graphs with error bars to find the uncertainty in the gradient and intercept
- combine the uncertainties in values to estimate the uncertainties in a calculated value.

The techniques for these skills are covered in the AS Study and Revision Guide, Sections 3.5 and 3.6.

P.1.3 Conclusions and evaluation

One of the first tasks following graph plotting is to consider whether it is consistent with the given relationship. This involves judging whether the points lie on a good straight line and possibly whether it passes through the origin. If you have plotted error bars, these involve deciding whether it is possible to draw a straight line through the error bars and whether the max/min graphs straddle the origin, respectively.

In addition, you will usually determine values of constants in the relationship by taking measurements from the graph – the gradient and intercept.

>> **Pointer**

Many physicists consider graphs to be purely numerical so the gradients and intercepts have no units. They would rephrase the Example: A graph of $(v/\text{m s}^{-1})^2$ against (x/m) is found to have a gradient of 1.50 ± 0.05 and intercept 18.2 ± 0.8.

Example

A graph of v^2 against x is found to be a straight line of gradient $1.50 \pm 0.05\,\text{m s}^{-2}$ and intercept $18.2 \pm 0.8\,\text{m}^2\text{s}^{-2}$. (See **Pointer**)

Calculate the acceleration, a, and initial velocity, u.

Answer

The relationship is $v^2 = u^2 + 2ax$

So the gradient is $2a$ and the intercept is u^2.

$\therefore$ Acceleration $= \dfrac{1.50 \pm 0.04}{2} = 0.75 \pm 0.02$ m s^{-2}

Fractional uncertainty in $u^2 = \dfrac{0.8}{18.2} = 0.044$

$\therefore$ Fractional uncertainty in $u = \frac{1}{2} \times 0.044 = 0.022$

$\therefore u = \sqrt{18.2}\,(1 \pm 0.022)\text{m s}^{-1} = 4.27 \pm 0.09$ m s^{-1}

quickfire

⑤ A graph of ln y against x is a straight line with gradient −0.15 and intercept 10 on the ln y axis. Determine the relationship between x and y.

>> **Pointer**

A risk assessment is **never** just a restatement of normal laboratory behaviour rules, such as keeping your bag under the bench. It must be specific to the procedure.

P.1.4 Risk assessment

Risk assessments form part of the NEA. They could also come up in the component tests. There are two kinds of risk assessment to consider.

(a) Maintaining the safety of the experimenter

This is generally performed under three headings:

- Identification of a **hazard**, e.g. burn hazard, slip hazard, cutting hazard.
- Identification of a **risk**, which is the specific aspect of the activity which brings the hazard into play, e.g. spilling hot water onto yourself when pouring water into a test tube.

Suggestion of a **control measure** to reduce the risk.

Example 1

Write a risk assessment for measuring the volume of a glass microscope slide, using a vernier calliper.

Answer

Hazard – cut hazard on broken glass.

Risk – cutting the skin when picking up the slide after a breakage, e.g. by dropping or by over-tightening the calliper jaws.

Control measures – care when holding; avoid hand contact with broken glass when clearing away broken slide.

Example 2

Write a risk assessment for connecting up an electron beam deflection tube to a UHT supply.

Hazard – shock hazard – this is a serious hazard.

Risk – bare leads of supply contacting the hand – medium risk.

Control measures – only connect leads when UHT supply turned off; use the output terminals of the supply which have a high ($M\Omega$) internal resistance to reduce the maximum current.

(b) Protecting the integrity of the apparatus

Many A level Physics procedures present negligible hazards to the person carrying them out. The risks to be assessed are therefore to the integrity of the apparatus, e.g. the possibility of breakage by overloading or over-extending a spring. You should plan to be aware of this possibility and avoid it. At least one significant risk should be identified together with a statement of how to avoid it. If no risks can be identified, this should be stated.

Example 1

Write a risk assessment for the investigation into the relationship between the period of oscillation of the loaded metre rule and the projecting length.

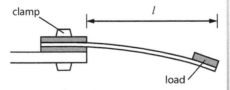

Fig. P.1 Loaded cantilever

Answer

A possible risk is that of the metre rule breaking. This is more likely to happen for large projecting distances, large amplitude oscillations and large loads. It can be avoided by careful increasing of the projecting distance and monitoring the deflection; also by restricting the amplitude of oscillation to a few cm.

quicKpire

⑥ 'For less than 1 minute of contact, currents greater than 4 mA are required to cause ventricular fibrillation.' In the light of this statement, comment on the maximum current obtainable from a 5 kV UHT supply with an internal resistance of 1 MΩ.

quicKpire

⑦ Even in the procedure in the example, it is possible to identify hazards, albeit trivial, to the experimenter. What can you spot?

P.2 Carrying out practical investigations

P.2.1 Planning

You will need to consider the apparatus available and how it can be used to carry out the investigation. In cases where you are not given detailed instructions, you will need to do some or all of the following:

- decide how to put the apparatus together to allow you to take results;
- assemble it and draw a diagram of your setup (note that diagrams of disassembled apparatus are neither required nor informative);
- take sufficient preliminary readings to allow you to establish ranges, intervals and the repeatability of the results (so that you can decide how many repeats are needed);
- write a brief plan outlining the decisions you have made including the nature of the graph you will draw to obtain a straight-line;

You will carry out the procedure in accordance with the instructions or your plan. If you decide to change the plan, you should include a note in your report as to the nature of the changes and the reasons for them.

P.2.2 Analysing results

Whether you are analysing your own results or using given data (e.g. in an exam question) the procedures are basically the same:

(a) The graph

Following drawing your graph and the first question to be answered is whether it is consistent with the expected relationship. Usually this means considering whether the data points are consistent with a straight line; whether the sign of the gradient (+ or -) is as expected; whether the data are consistent with a proportional relationship. If error bars are drawn (non-log graphs) the test for a straight line involves noting whether such a line can be drawn through them. The method of handling error bars in graph plotting is covered in the AS SRG.

(b) Measurements from the graph

Usually this involves determining the gradient and intercept together with an estimate of the uncertainty in these values. Either of these is then normally used to determine the value of a physical quantity.

Example

In an oscillating spring experiment, the period of oscillation, T, is determined for a range of suspended masses, m. The relationship is expected to be:

$$T = 2\pi\sqrt{\frac{m}{k}}$$

What graph would you draw to verify the relationship and how would you use it to determine a value for the spring constant, k?

Answer

A graph of T^2 against m can be drawn. The relationship is supported if the data points are consistent with a straight line through the origin. The gradient is equal to $\dfrac{4\pi^2}{k}$ so to determine k the gradient should be measured and then $k = \dfrac{4\pi^2}{\text{gradient}}$.

quickⲫire

⑨ The value of T is measured as $1.53 \pm 0.02\,\text{s}$. What is the total length of the error bar in T^2?

❱❱ Pointer

An alternative graph which can be plotted is T against $\sqrt{m}$. See **extra questions**.

extra

1. A student obtains a set of readings of period, T, for a series of masses, m, on an oscillating spring. He plots a graph of (T/s) against $\sqrt{(m/\text{kg})}$, obtains a straight line through the origin and determines the gradient to be 1.18 ± 0.04. Given the relationship, $T = 2\pi\sqrt{\dfrac{m}{k}}$, determine a value for k together with its absolute uncertainty.

2. A student uses a metre rule (resolution $1\,\text{mm}$) and a laboratory balance ($0.01\,\text{g}$) to determine the density of the glass of geological microscope slides. Her results are as follows:

   Length of 10 microscope slides laid end to end = $46.2\,\text{cm}$

   Width of 10 microscope slides laid side to side = $26.8\,\text{cm}$

   Thickness of 30 stacked microscope slides = $3.05\,\text{cm}$

   Mass of 30 microscope slides = $95.57\,\text{g}$

   (a) Without doing any calculation, state, with a reason, which measurement makes the greatest contribution and which makes the least contribution to the uncertainty in the calculated density.

   (b) Calculate the percentage uncertainty in each of the measurements.

   (c) Use the measurements to determine the density of the glass, together with its absolute uncertainty.

3.   In an experiment to measure the internal resistance of a battery, a student connects three 10 Ω resistors, with a stated tolerance of 1% (a) in series and (b) in parallel.

For each combination, calculate the maximum and minimum resistance and hence give the resistance of the combination together with its absolute uncertainty.

4.   The student in Q3 connected the resistors, singly and in combination, across a battery of two D-type cells and measured the terminal pd each time. The results were as follows:

External resistance, R / Ω	3.33	5.00	6.67	10.0	15.0	20.0	30.0
Terminal pd, V / V	2.78	2.90	2.96	3.02	3.07	3.09	3.11

(a) Show by sketches how the 6.67 Ω and 15 Ω resistances were achieved.

(b) Determine the internal resistance and emf from a plot of terminal pd against current ($V = E - Ir$).

(c) Show that the terminal pd can be calculated from the external resistance using the equation $\dfrac{1}{V} = \dfrac{1}{E} + \dfrac{r}{ER}$.

(d) By plotting a suitable graph, use the result of part (c) to determine the internal resistance and emf.

(e) The student's teacher suspected the student of cheating. What is the evidence?

5.   Two students use a digital centisecond stopwatch to measure the period, T, of the same simple pendulum.

Student **A** measures the time for 5 oscillations four times. His readings are 4.35 s, 4.31 s, 4.38 s and 4.28 s.

Student **B** measures the time for 20 oscillations once. Her reading is 17.28 s.

(a) Give Student A's result for T together with its absolute uncertainty.

(b) Assuming that the uncertainty for Student B's timings is the same as for Student A's, give Student B's result for T together with its absolute uncertainty.

(c) Evaluate the two methods for determining T.

6.   A group of students investigates the how the depression, y, of a loaded metre rule depends upon the separation, l, of the supports. The setup is shown in Fig. P.2. The load is at the centre and the supports are symmetrically placed. The depression, y, is measured using a travelling microscope which is focused on the central point, **P**.

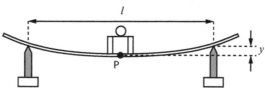

Fig. P.2 Loaded beam

Results:

l / cm	40.0	45.0	50.0	56.0	60.0	66.0	70.0	76.0	80.0
y / mm	0.85	1.14	1.61	2.38	2.75	3.74	4.42	5.73	6.63

The students expect that the relationship between the l and y is $y = Al^n$, where A and n are unknown constants. Plot an appropriate graph to verify this and to determine the values of A and n.

7. The students in Q6, read that the deflection of the beam is given by:

$y = \dfrac{mgl^3}{4Eab^3}$, where m = load mass, E = the Young modulus of the material of the rule

a = width of the metre rule, b = thickness of the metre rule

(a) Compare this relationship to the result of Q6.

(b) The students make the following measurements:

$m = 200 \pm 0.01\,\text{g}$

$a = 2.80 \pm 0.01\,\text{cm}$

$b = 4.50 \pm 0.01\,\text{mm}$

They estimate the uncertainty in the y values as $\pm 0.05\,\text{mm}$ and assume the uncertainty in the l measurements is $\pm 1\,\text{mm}$.

(i) Suggest an appropriate graph to plot to determine E along with its absolute uncertainty. Explain how you will find E.

(ii) Plot the results with appropriate error bars and draw the lines of maximum and minimum gradient.

(iii) Determine the gradient together with its absolute uncertainty

(iv) Determine the value of E together with its absolute uncertainty.

(v) Identify the main source of random uncertainty in the experiment.

M Mathematics and data

The A level qualification requires you to manipulate more complex mathematical relationships than AS and to analyse data sets which are related by such relationships. This section builds upon the Section 3 of Year 1 & AS Study and Revision Guide.

M.1 Graphs and error bars with non-linear functions

》 Pointer

Remember that 'y against x' means that y is on the vertical axis and x on the horizontal.

You are expected to plot linear graphs for non-linear relationships, which could be more complicated than those at AS level, and also to include error bars. This section is essentially a brief recap on the AS book with a more complicated example.

M.1.1 Linearising graphs

Remember that the trick is to convert the equation into the form $y = mx + c$, where x and y are the variables. Then a graph of y against x has a gradient m and an intercept c on the y axis.

As an example, consider Fig. M1. A *compound pendulum* is a freely suspended distributed mass. In this case the mass is a uniform bar. Using the symbol on the diagram, the period, T, of the oscillation is given by:

$$T = 2\pi \sqrt{\frac{k^2 + l^2}{gl}}$$

where k is a constant and g the acceleration due to gravity. We'll use this as an example of a more complex (but still algebraic) relationship which we need to manipulate. This is as complicated as it gets!

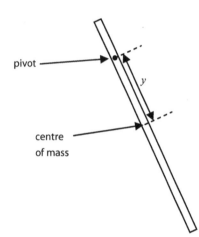

pivot

centre of mass

Fig. M1 Compound pendulum

Example
The period of the compound pendulum in Fig. M1 is measured for a range of values of l, in order to verify the relationship above. What plot will give a straight line?

Answer

Squaring gives $\qquad T^2 = 4\pi^2 \left(\dfrac{k^2 + l^2}{gl} \right)$

Rearranging gives $\qquad T^2 l = \dfrac{4\pi^2 k^2}{g} + \dfrac{4\pi^2}{g} l^2$

Comparing this with the linear equation $y = mx + c$ suggests that a graph of $T^2 l$ against l^2 should be linear.

quicKfire

① Suggest a unit for k in the compound pendulum equation. Explain your answer.

quicKfire

② For the compound pendulum graph, state:
a) the gradient
b) intercept on the $T^2 l$ axis.

M.1.2 Error bars

We use the rules for combining uncertainties (see **Pointer**) to calculate the lengths of the error bars.

Example

During an investigation of a compound pendulum, as in Fig. M1, a student obtained the following results:

$$l = 30.0 \pm 0.2\,\text{cm} \qquad T = 2.81 \pm 0.05\,\text{s}$$

Determine the horizontal and vertical error bars for this point when plotting a graph of T^2l against l^2.

Answer

Fractional uncertainties: $p(l) = \dfrac{0.2}{30.0} = 0.0067$; $p(T) = \dfrac{0.05}{2.81} = 0.018$

Using the rules for combining uncertainties:

$$p(l^2) = 2p(l) = 0.013;$$

and

$$p(T^2l) = 2p(T) + p(l) = 0.036 + 0.067 = 0.043$$

$T^2l = (2.81\,\text{s})^2 \times 30.0\,\text{cm} = 237\,\text{s}^2\,\text{cm}$ and $l^2 = 900\,\text{cm}^2$, so the absolute uncertainties in T^2l and l^2 are given by:

$$\Delta(l^2) = l^2 \times 3p(l^2) = 900\,\text{cm}^2 \times 0.013 = 12\,\text{cm}^2$$

and $\qquad \Delta(T^2l) = T^2l \times p(T^2l) = 237\,\text{s}^2\,\text{cm} \times 0.043 = 10\,\text{s}^2\,\text{cm}$

The error bars are illustrated in Fig. M2.

M.2 The exponential and logarithm functions

The exponential function is one of the most useful mathematical functions in physics. The number, e, also known as Euler's number, has a value of 2.718... to 4 sf but, like π, it is an irrational number.

Its main use in A level Physics relates to decays which have a constant half-life, as in Fig. M3. These are referred to as *exponential decays*. Examples include:

- capacitor discharge
- radioactive decay
- damped motion
- γ-ray absorption.

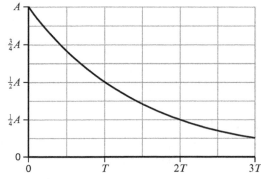

Fig. M3 Exponential decay

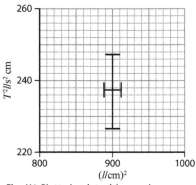

Fig. M2 Plotted point with error bars

quickfire

③ With $l = 20$ cm, the period, T, of the student's compound pendulum, in the example was found to be 3.30 s. Determine the values of T^2l against l^2 together with their absolute uncertainties.

>> *Pointer*

The number, e, is an irrational number (i.e. like π, it cannot be expressed as a ratio of two integers) and has a value of 2.718 (to 4 sf).

Alongside the trig functions (Section M.4), these are the most useful functions used in A level Physics.

M.2.1 The exponential growth function, e^x

For $a > 1$, the graphs of $y = a^x$, all look very similar – see Fig. M4. They all pass through the point (0, 1) and they all get steeper and steeper. In fact, for each of them, the gradient is proportional to their value (i.e. when x has a value such that $a^x = 20$, the gradient is twice as much as when $a^x = 10$).

There is one special number, e, such that the gradient of $y = e^x$ is *equal* to the value (e^x). The number e (see Pointer) has a value (to 4 sf) of 2.718 and is called Euler's number after the Swiss mathematician. The graph in Fig. M5 is of $y = e^x$ for values of x between −2 and +3 (though we shall be interested in the positive values of x).

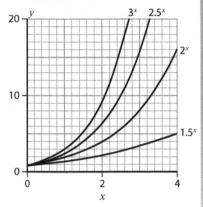

Fig. M4 $y = a^x$ graphs

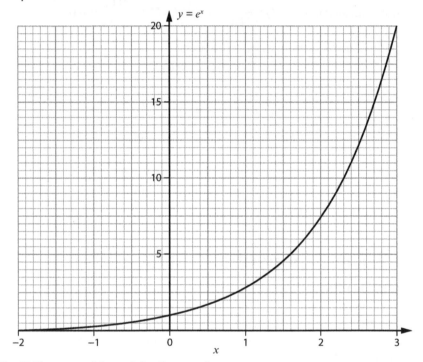

Fig. M5 The exponential growth function, $y = e^x$

CALCULATOR TIP

The value of e^x is found using the e^x key. On some calculators, this is the same as the ln key but accessed using the SHIFT key. Check this by finding e^{10} (148.4).

Because it grows at a rate which is equal to its magnitude, $y = e^x$ is called the exponential function or the growth function. So, for example, when $e^x = 10$, the gradient is also equal to 10. One of the consequences of this is that the function increases by equal proportions in equal intervals of x. Thus we can use e^x to model systems which grow at a rate proportional to their size.

quicKpire

④ Use your calculator to determine $e^{-1.5}$, $e^{-0.5}$, $e^{0.5}$, $e^{1.5}$ and $e^{2.5}$.

Example

Show that the exponential function grows by the same factor every time x increases by 1 and find that factor.

Answer

Using the rules of indices, $\dfrac{e^{x+1}}{e^x} = e^{(x+1)-x} = e^1 = e$

Thus, whatever the value of x, e^{x+1} is e times as big as e^x, i.e. approximately $2.718 \times$ as big.

Generally, physical systems can only grow according to the exponential growth function for a restricted time before the system becomes self-limiting (e.g. bacterial colonies which run out of food, a nuclear bomb which runs out of fuel).

M.2.2 The exponential decay function, e^{-x}

There are many physical quantities which become *smaller* at a rate which is proportional to their size. These can be usefully modelled using the *exponential decay function*, $y = e^{-x}$, which is plotted in Fig. M6 for x between 0 and 2.5.

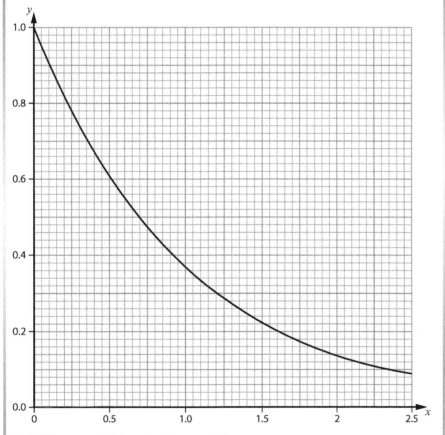

Fig. M6 The exponential decay function, $y = Ae^{-x}$

Examples of such systems include radioactive decay, capacitor discharge, γ-ray absorption and damped motion. These are all dealt with in Section M.3

Rules of indices

For any value of a apart from the exceptions given:

1. $a^{m+n} = a^m \times a^n$

2. $\dfrac{1}{a^m} = a^{-m} \ (a \neq 0)$

3. $a^{m-n} = \dfrac{a^m}{a^n} \ (a \neq 0)$

4. $a^0 = 1 \ (a \neq 0)$

5. $\sqrt[n]{a} = a^{\frac{1}{n}} \ (a \geq 0$ for even values of $n)$

≫ Pointer

The graph of $y = e^{-x}$ is the reflection of the $y = e^x$ graph in the y axis. Note the different vertical scales of the two graphs in Fig. M5 and Fig. M6.

quickpire

⑤ Show that, if x increases by Δx, the exponential decay function, e^{-x}, decreases by a factor which does not depend on x. Determine this factor for $\Delta x = 0.5$.

» Pointer

'Logarithm' is usually shortened to 'log', so ln x is the natural log of x or just 'log x'. The process of finding ln x is called 'taking logs'.

» Pointer

Let $e^x = 100$. From Fig. M4 we know that $e^{2.3} \sim 10$
So $e^{2.3} \times e^{2.3} \sim 100$
Now, by using the calculator, $e^{4.6} = 99.5$, so try $e^{4.61}$. This gives 100.5, so we'll try $e^{4.605}$ which gives 99.98. We'll stop there.

quickfire

⑥ Solve the following equation for t:
$800e^{-25t} = 10$.

quickfire

⑦ We know that any number (other than 0) to the power of 0 is 1. So, in particular, $e^0 = 1$.

Deduce the value of ln1.

» Pointer

These six properties of logs are not really all different. For example, you should be able to derive 3 and 4 from 1 and 2.
[Hint: Use the answer to Quickfire 7.]

quickfire

⑧ Write equivalent properties for the $\log_{10}$ function.

M.2.3 The logarithm function, ln x

Suppose we know that $e^x = 100$. What is the value of x? We could find it by using Fig. M5 and a bit of trial and error (see **Pointer**), which gives $x = 4.605$ (to 4 sf). However, there is an easier way.

(a) The definition of the natural logarithm function, ln x

The *natural logarithm* function, ln, is defined as the **inverse function** to the exponential function. In other words

$$\ln(e^x) = x$$

The ln button on the calculator (which is often the same button as the e^x button) calculates this function and we can use it to answer the question which we posed above:

If $e^x = 100$ then $\ln(e^x) = 100$

But, by definition, $\ln(e^x) = x$.

$\therefore x = \ln 100 = 4.6052$ (to 5 sf) (which agrees with our trial and error approach!)

The process of finding the logarithm is often referred to as 'taking logs'.

Example

Solve the equation $10e^{-2x} = 3$

Answer

Dividing the equation by 10: $e^{-2x} = 0.3$
Taking logs: $\ln(e^{-2x}) = \ln 0.3 = -1.204$ (to 4 sf)
$\therefore -2x = -1.204$, so $x = 0.602$ (to 3 sf)

(b) Logs to other bases

Natural logarithms are sometimes referred to as 'logs to the base e' because ln x is the inverse function to e^x. It is possible to define log functions to any positive base but the only one in common use is 10.

The logarithm to the base 10, written $\log_{10}x$, is defined as the inverse function to 10^x, i.e. $\log_{10}10^x = x$.

(c) Properties of logs

In order to solve equations using logs, you need to know the following properties. They can be derived using the definition of the log function[2].

1. $\ln e^x = x$ (the definition)
2. $\ln ab = \ln a + \ln b$
3. $\ln\dfrac{1}{a} = -\ln a$ (See **Pointer**)
4. $\ln\dfrac{a}{b} = \ln a - \ln b$
5. $\ln a^n = n \ln a$
6. $e^{\ln x} = x$

[2]See Maths for A level Physics (2016) by Kelly and Wood, Section 4.4.3

M.2.4 Modelling systems using the exponential function

Most of the systems we shall model are time decays, so we shall use time, t, as the independent variable. The graphs in Fig. M7 are of the function $x = Ae^{-kt}$ for three different values of k. When $t = 0$, the value of e^{-kt} is 1, so all the graphs pass through $(0, A)$.

The larger the value of k the more rapidly the graphs decay. We use this to model different decays by observing the following properties in sequence:

- The gradient of $x = e^{-t}$ is equal to minus the value of e^{-t} (check that the gradient of of the $k = 1$ graph in Fig. M7 is $-\frac{1}{2}A$ where it crosses the $x = \frac{1}{2}A$ line). [See first **Pointer**].

- The gradient of $x = Ae^{-t}$ is equal to $x = -Ae^{-t}$, so again the gradient of is equal to minus the value of the function.

- The gradient of $x = Ae^{-kt}$ is equal to $x = -Ake^{-kt}$ (use Fig. M7 to check that the gradients for $k = 2$ is twice that for $k = 1$ which is twice that for $k = 0.5$).

> Those familiar with calculus can replace the above bullet points with the rather snappier:
>
> If $\quad x = Ae^{-kt}$
>
> Then $\quad \dfrac{dx}{dt} = -Ake^{-kt} = -kx$

Hence if we encounter a physical system, the rate of change of which is proportional to its instantaneous value, we can immediately write its decay equation. Here are the examples you will meet in A level Physics.

(a) Radioactive decay

The rate of decay, i.e. the activity, A, of a radioactive sample consisting of a single nuclide is proportional to the number of atoms, N, in the sample, i.e. $A = \lambda N$ where λ is a constant called the *decay constant*.

But the activity is just the rate of decrease of N, i.e. $A = -\dfrac{dN}{dt}$.

So $\quad \dfrac{dN}{dt} = -\lambda N \qquad$ Hence $\quad N = N_0 e^{-\lambda t}$

where N_0 is the number of radioactive atoms at time $t = 0$.
But A is proportional to N so it is also true that $A = A_0 e^{-\lambda t}$.

(b) Damped motion

The resistive force on bodies moving slowly through a medium is proportional to the velocity, i.e. $F = -\alpha v$. Hence, for a body of mass m, the velocity decreases according to $A = A_0 e^{-kt}$ where $k = \dfrac{\alpha}{2m}$.

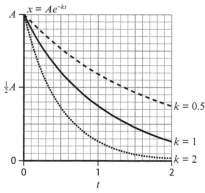

Fig. M7 $x = Ae^{-kt}$ graphs

>> **Pointer**

You can do the 'checks' in bullet points 1 and 3 by drawing a larger version of the graphs in Fig. M7 and constructing tangents.

>> **Pointer**

The decay constant, λ, has dimensions of time^{-1}, i.e. if t is in seconds, the unit of λ is s^{-1}. This must be the case because the value of $e^{-\lambda t}$ at any time cannot depend upon what units we are using, i.e. λt must have no units.

quickfire

⑨ A damped oscillation has a decay constant of $0.1\,\text{s}^{-1}$.

How long does it take for the amplitude to fall from 8 cm to 1 cm?

The analysis of damped harmonic motion requires second order differential equations. This leads to the equation[3]

$$x = A_0 e^{-kt} \cos(\omega t + \varepsilon)$$

In other words, the amplitude of the oscillation decays according to $A = A_0 e^{-kt}$.

(c) The absorption of X rays and γ rays

If a beam of X or γ rays passes through a material, the probability of a photon being absorbed in passing through a given thickness Δx of material is constant. Hence the number being absorbed in any section Δx is proportional to the number of photons present.

The intensity, I, is proportional to the number of photons per second, so $I = I_0 e^{-\mu x}$ where μ is the absorption coefficient.

M.2.5 Time to decay

If a quantity, x, decays according to $x = x_0 e^{-\lambda t}$, then it approaches zero as time increases without ever reaching it. However, we can find the time to decay to a certain fraction of the original value. This is normally done in one of two ways.

(a) The decay time

The quantity $\frac{1}{\lambda}$ is defined as the decay time. If $t = \frac{1}{\lambda}$, then $e^{-\lambda t} = e^{-1}$. Hence in this time, the amplitude falls to e^{-1} of its original value, or approximately 0.37 (37%). This fact can be used as a quick way of determining an approximate value of λ from an experimental decay graph.

Example

Engineers frequently use five times the decay time as being effectively the time for a complete decay. Show that in this time the amplitude is reduced by more than 99%.

Answer

If $t = \frac{5}{\lambda}$, $e^{-\lambda t} = e^{-5} = 0.0067$ (to 2 sf)

0.0067 is less than 0.01 which is 1%. Hence the amplitude has decayed by more than 99%.

(b) The half-life

This quantity can be applied to any exponential decay but it is only commonly used in radioactive decays.

We can derive an expression for the half-life by inserting the value $x = \frac{x_0}{2}$ into the equation $x = x_0 e^{-\lambda t}$.

Then $\frac{x_0}{2} = x_0 e^{-\lambda t}$, which simplifies to $\frac{1}{2} = e^{-\lambda t}$ which we can rewrite as $e^{\lambda t} = 2$.

[3] See Maths for A level Physics (2016) by Kelly and Wood.

If we 'take logs' of this equation, we get $\ln(e^{\lambda t}) = \ln 2$.

But $\ln(e^{\lambda t}) = \lambda t$, $\quad \therefore \lambda t = \ln 2$, $\quad$ i.e. $t = \dfrac{\ln 2}{\lambda} \approx \dfrac{0.69}{\lambda}$

Notice that the value of the half-life is about 70% that of the decay time.

M.3 Analysing data using log graphs

You will be expected to use log graphs in data analysis in both your theory exams and your practical work.

M.3.1 Power law relationships

Relationships of the form $y = Ax^n$, where A and n are constants, are called power law relationships. If A and n are unknown, we can use logs to:

- verify that the relationship is indeed of this form, and
- determine the values of A and n.

If $y = Ax^n$, then taking logs of both sides gives $\ln y = \ln(Ax^n)$

Using the properties of logs in Section M.2.3(c), we can rewrite this as:

$$\ln y = \ln A + \ln x^n$$

and further $\qquad\qquad\qquad \ln y = \ln A + n\ln x$

Re-ordering for convenience $\qquad \ln y = \boxed{n}\ln x + \boxed{\ln A}$

Comparing with the straight line equation: $\quad y = \boxed{m}\,x\; +\; \boxed{c}$

Hence, a graph of $\ln y$ against $\ln x$ is a straight line with a gradient n and an intercept of $\ln A$ on the $\ln y$ axis. In order to determine the value of A, we use the inverse relationship between the logarithm and exponential functions, so $A = e^{\ln A} = e^{\text{intercept}}$.

Example

A student investigating a power law relationship between two variables, v and s, plots a graph of $\ln v$ against $\ln s$, and finds that the gradient is 2.5 and the intercept on the $\ln v$ axis is -1.36. Find the relationship.

Answer

If $v = As^n$, then the gradient of the graph is n. Hence $n = 2.5$

The intercept on the $\ln v$ axis is $\ln A$. $\therefore \ln A = -1.36$

$\therefore A = e^{-1.36} = 0.26$ (to 2 sf)

$\therefore v = 0.26\, s^{2.5}$

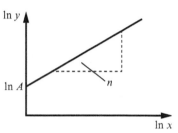

Fig. M8 A log–log plot

quickfire

⑬ A graph of $\ln y$ against $\ln x$ is a straight line of gradient 2 and an intercept of 2.5 on the $\ln y$ axis. What is the relationship between x and y?

quickfire

⑭ The luminosity, L, of stars of mass M, such that $2M_\odot < M < 20M_\odot$, where $M_\odot$ is the solar mass, is given approximately by

$$\frac{L}{L_\odot} = 1.5\left(\frac{M}{M_\odot}\right)^{3.5}.$$

How could this relationship be verified?

quickfire

⑮ The activity, A, of a radioactive source varies with time according to $A = A_0 e^{-\lambda t}$, where λ is the decay constant.

To determine λ what graph should be plotted and how will λ be found?

UNITS

When analysing vibrations, angles are always expressed in radians. You should make sure your calculator is in the appropriate mode.

Check: $\cos \pi = -1$.

M.3.2 Exponential relationships

Relationships of the form $x = Ae^{-kt}$, where A and k are constants, can be verified by taking logs and the values of A and k found.

If $x = Ae^{-kt}$, then taking logs of both sides gives $\ln x = \ln(Ae^{-kt})$

∴ Using the properties of logs: $\ln x = \ln A + \ln e^{-kt}$

∴ $\ln x = \boxed{-k}\, t + \boxed{\ln A}$

$\hspace{3.5cm} \updownarrow \hspace{1.6cm} \updownarrow$

Comparing with the straight line equation: $y = \boxed{m}\, x + \boxed{c}$

So a graph of $\ln x$ against t is a straight line with gradient k and an intercept of $\ln A$ on the $\ln x$ axis.

M.3.3 Error bars and log plots

You will not be expected to use error bars when you undertake log–log or semi-log plots.

M.4 Trigonometrical functions

In addition to using trig functions to solve triangles, e.g. in resultant vector calculations, this part of the course requires their use in vibrations (and in the AC electricity option).

M.4.1 Basic trig graphs

You are required to recognise the following trig graphs and to use them in modelling physical relationships.

(a) sin x and sin²x

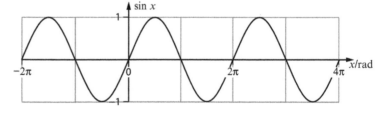

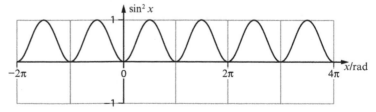

Fig. M9 The relationship between $\sin x$ and $\sin^2 x$

Things to notice about the $\sin^2 x$ function:

1 It is always positive.

2 It is a sinusoid – notice the shape at the bottom of the curves where $x = -\frac{1}{2}\pi, \frac{1}{2}\pi, \frac{3}{2}\pi \ldots$

3 The mean value is $\frac{1}{2}$.

4 $\sin^2(-x) = \sin^2 x$

(b) $\cos x$ and $\cos^2 x$

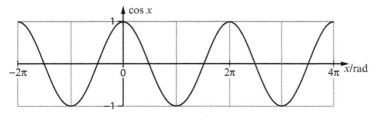

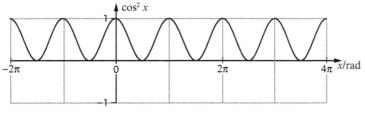

Fig. M10 The relationship between $\cos x$ and $\cos^2 x$

M.4.2 Handling A$\cos(\omega t + \varepsilon)$

Simple harmonic oscillations are described using $x = A\cos(\omega t + \varepsilon)$, where x is the displacement, A the amplitude and ω the angular frequency or pulsatance. It can also be written $x = A\cos(2\pi f t + \varepsilon)$, where f is the frequency.

With $\qquad\qquad x = A\cos(\omega t + \varepsilon)$

the velocity $\qquad v = -A\omega\sin(\omega t + \varepsilon)$

and the acceleration $\qquad v = -A\omega^2\cos(\omega t + \varepsilon)$,

the effect of the phase constant, ε, is to shift the cosine and sine graphs to the **left** (if ε is positive) by a time $\dfrac{\varepsilon}{\omega}$. The form of the graph of x against t is shown in Fig. M11.

>> *Pointer*

$\sin^2$ and $\cos^2$ functions are useful when plotting the variation of potential and kinetic energy with time for an object undergoing simple harmonic motion.

Grade boost

You can use trig identities to show that
$\sin^2 x = \frac{1}{2} - \frac{1}{2}\cos 2x$
and hence that $\langle \sin^{-1} x \rangle = \frac{1}{2}$
and also that $\langle \cos^{-1} x \rangle = \frac{1}{2}$.

quickfire

⑯ Use your calculator to find $\cos 0.5$ rad, $\cos^2 0.5$ rad, $\sin 0.5$ rad, $\sin^2 0.5$ rad, and then without further calculation state the values of $\cos(-0.5\text{ rad})$, $\cos^2(-0.5\text{ rad})$, $\sin(-0.5\text{ rad})$, $\sin^2(-0.5\text{ rad})$, [Hint: consider the symmetry of the graphs.]

>> *Pointer*

The specification uses the $A\cos(\omega t + \varepsilon)$ form of the oscillation function. Use of $A\sin(\omega t + \varepsilon)$ is equally acceptable.

quickfire

⑰ Sketch graphs of the functions:

   a) $10\cos(8\pi t + 0.1\pi)$

   b) $4\cos(20t - 1)$.

quickfire

⑱ Find the least two values of $t > 0$ for which the function $x = 5\cos(10t + 1)$ has the value -5.

quickfire

⑲ Use $v = -A\omega\sin(\omega t + \varepsilon)$ to find the velocity of the particle at the two times

$$t = \frac{2\pi \pm (0.927 - 1)}{10}$$

and hence show that the answer in the example is correct.

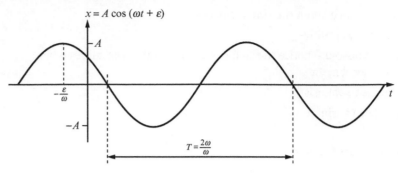

Fig. M11 The form of the function $x = A\cos(\omega t + \varepsilon)$

Finding the times when the displacement is a maximum (positive or negative) is straightforward. For other displacements, it is more complicated.

Example

A particle oscillates with $x = 5\cos(10t + 1)$, with x in cm and t in s. Calculate the first time after $t = 0$ when $x = 3$ cm and the velocity is positive.

Answer

If $5\cos(10t + 1)$, then $\cos(10t + 1) = 0.6$, $\therefore 10t + 1 = \cos^{-1} 0.6 \dots$

The calculator gives $\cos^{-1} 0.6 = 0.927$. From the cosine graph, -0.927 is also a solution as well as $2\pi \pm 0.927$ etc. Looking at the lowest positive answers gives

$$t = \frac{2\pi \pm 0.927 - 1}{10},$$ leading to 0.436 s as the lowest answer (see Quickfire 19).

1. Kepler's third law states that the square of the period, T, of a planet's orbit is proportional to the cube of a, the semi-major axis. Give two non-log plots which would produce straight-line graphs.

2. Applying Newton's laws to the motion of satellites in circular orbits gives the following relationship between the period, T, and radius, r:

$$T = 2\pi \sqrt{\frac{r^3}{GM}}.$$

The mass of Uranus is 8.68×10^{25} kg. Find the gradient and intercept for the following graphs for the satellites of Uranus:

   (a) $\ln (T/s)$ against $\ln (r/m)$
   (b) $(T/s)^2$ against $(r/m)^3$
   (c) (T/day) against $(r/10^6 \text{ km})^{\frac{3}{2}}$

3. A particle oscillates with displacement x (in cm) related to time t (in s) according to the equation $x = 5.0\cos(6\pi t + 1.0)$.

   (a) State the amplitude, frequency, period and angular frequency.
   (b) State the position at $t = 0$.
   (c) Calculate (i) the position, (ii) the velocity and (iii) the acceleration when $t = 1.4$ s.
   (d) Calculate the speed of the particle when $x = 4.0$ cm.

4. The radioactive tracer, iodine123 has a half-life of 13.22 hours. Calculate:

   (a) the value of the decay constant, λ
   (b) the activity of a 1.00 nmol sample of iodine123 (i.e. 1.0×10^{-9} mol)
   $[N_A = 6.02 \times 10^{23}\,\text{mol}^{-1}]$
   (c) the activity and number of atoms of iodine123 remaining from the 1 nmol sample after 1 week.

5. A particle of mass 0.20 kg oscillates with amplitude 10 cm and period 0.25 s. It is at its maximum displacement when $t = 0$.

   (a) Write down the position of the particle as a function of time.
   (b) Write down the velocity of the particle as a function of time.
   (c) Calculate the maximum kinetic energy of the particle.
   (d) Taking the minimum potential energy of the particle to be zero, sketch graphs of the variation of kinetic energy and potential energy of the particle between 0.0 and 0.5 s.

6. Two radioactive nuclides, **A** and **B**, have half-lives of 3 and 6 days respectively. The initial activity of **A** is 4 times that of **B**. After how many days will their activities be equal? What fraction of each remains?

Exam practice and technique

Answering exam questions

Some students never ask themselves how a question is put together or the sort of skills they will need to demonstrate in order to achieve good marks. This is a mistake. You'll expect to gain more marks if you are able to read the question setter's mind when looking at a question and understand the sort of answer which the examiner is expecting (or at least, hoping for).

When setting questions, examiners work within certain rules so that an exam paper one year tests the same sort of abilities as in other years – without using the same questions! The main constraints on questions are called 'Assessment Objectives' (AOs)

Assessment objectives

AO1: 30% of the marks are for showing that you recall and understand aspects of physics. For example, you can state a law or definition, you know what equation to use to solve a problem or you can describe how you would carry out an experiment.

AO2: 45% of the marks are for using the AO1 knowledge in order to solve problems. This involves producing answers to calculations, bringing together ideas to explain things, combining and manipulating formulae, using experimental results and graphs.

AO3: 25% of the marks are for such things as reaching conclusions from experimental results or other data, for designing experiments or refining experimental techniques.

Skills

At the same time as balancing the AOs, the examiner looks at the balance of skills. Being physics, a high percentage of marks (at least 40%) comes from the application of **mathematics**.

This might actually seem a little low but the low-level skill of putting numbers into very basic equations doesn't count as a mathematical skill but is essentially just communication! The same is true of plotting graphical points.

On the other hand, drawing tangents and finding their gradients is certainly maths. So are rearranging equations and producing answers to numerical questions. In any case, examiners are free to demand more than 40% maths skills, without being excessive.

Questions involving designing, analysing and drawing conclusions from **experiments**, e.g. from the specified practical work, must account for at least 15%.

As an illustration, look at this question part:

The minimum potential difference, V, applied to a light emitting diode (LED) for it to be seen to emit light is related to the wavelength, λ, of the light by the approximate equation:

$$eV = \frac{hc}{\lambda}.$$

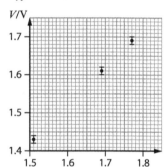

V is measured for three LEDs, and a graph of V against λ is plotted using values of λ supplied by the makers of the diodes.

(i) Calculate the maximum and minimum gradients of the graph. [2]

(ii) Hence calculate a value for the Planck constant, as well as its **percentage** uncertainty. [3]

(iii) Discuss whether or not the graph confirms the equation. [3]

There are no AO1 marks within this question because, although all parts require knowledge to answer them, you only start to accrue credit when you do something with the knowledge, e.g. you need to apply judgement to draw the max/min lines before you can determine their gradients. Within this 8-mark question part there are: 3 AO2 marks, 5 AO3 marks, 5 maths marks and all the marks (8) count as practical because they are in the context of a practical activity. The maths marks are for parts (i) and (ii). The AO3 marks are for parts (ii) [2 out of the 3 marks] and (iii).

Exam tips

Now we'll have a look at some tips to enable you to show what you do in answering exam questions. The first and most important point is to read the question carefully. Examiners discuss the wording of questions so that the meaning is clear and precise. In spite of this, it is easy to misinterpret a question, so take your time. Using a highlighter to mark key information often helps, e.g. numerical information given at the beginning of a question is sometimes not needed until later on, and so highlighting this makes it stand out.

Look at the mark allocation

Each part of a question is allocated a number of marks. In written answers, this total gives a hint as to how much detail you need in your answer. In calculations, some marks will be for the working and some for the answer [see below].

Understand the command words

These are the words which show the sort of answer the examiner expects in order to give you credit.

State

A short answer with no explanation.

Explain

Give a reason or reasons. Look at the mark allocation: 2 marks usually means that you need to make two distinct points.

State and explain

There may be a mark for the statement but the first mark may be for an explanation of a correct statement, e.g. *'State which resistor, A or B, has the higher value and explain your reasoning.'* It is unlikely that the examiner will give you a mark for a 50/50 choice!

Calculate

A correct answer will score all the marks, unless the question includes the instruction to *'show your working'*. **Warning**: An incorrect answer without working will score 0.

Always give the units of your answer – missing or incorrect units will be penalised.

Determine

This is an alternative to 'calculate' and examiners often use it when the method of calculation is more obscure e.g. (on a curved $v - t$ graph) determine the acceleration at 2.0 s.

Show that [in a calculation question]

E.g. 'Show that the resistivity is approximately $2 \times 10^{-7}\ \Omega$ m.' There is no mark here for the correct answer; the working must be shown in sufficient detail for the examiner to be convinced you know what you are doing! **Hint**: In this case, calculate an accurate answer, e.g. $1.85 \times 10^{-7}\ \Omega$ m and say that this is approximately the value stated.

Describe

A series of statements is required. These may be independently marked but care may be needed with sequencing, e.g. in the description of how to carry out an experiment.

Compare

There must be a clear comparison, not just two separate statements. It is also not safe just to state one thing and leave the examiner to infer another; e.g. *'Compare the work functions of metals A and B.'*

Answer 1: Metal A has a low work function – not enough.

Answer 2: Metal A has a **lower** work function **than metal B** – this answer would gain credit (if correct!) unless the question makes it clear that a numerical comparison is required.

Suggest

This command word often comes at the end of a question. You are expected to put forward a sensible idea based upon your physics knowledge and the information in the question. There will often not be a single correct answer.

Justify

Explain why a result or argument is correct. It often requires calculation.

Name

A single word or phrase is expected; e.g. 'State the name of the property of light being demonstrated' (*in a question showing waves spreading out after passing through a gap*). Answer: *Diffraction*. Note that, especially in this kind of question, a correctly spelt answer may be required.

Estimate

This does not mean 'guess'. It usually involves one or more calculations with simplifying assumptions. The question may ask you to state any assumptions you make. E.g. *Estimate the number of 1 mm diameter spheres which will fill a measuring cylinder up to the 100 cm³ mark.*

Derive

This involves producing a given equation starting from a set of assumptions and/or more basic equations. An example is using Newton's laws to derive Kepler's third law for the case of circular orbits, $T^2 \propto a^3$. Another example which you should know is the derivation of the relationship between the half-life, T, and the decay constant, λ, for a radioactive nuclide: $\lambda = \dfrac{\ln 2}{T}$.
Read the specification and make sure you know which derivations are required.

Tips about diagrams

Questions about experiments sometimes ask for diagrams. The diagram should show the arrangement of the apparatus and be labelled. Separate diagrams of a capacitor, a resistor and a voltmeter will gain no credit. Note, however, that standard circuit symbols, e.g. a cell or a voltmeter, do not need labelling. Even if the question does not demand one, some of the marks may be awarded for information included in a well-drawn diagram.

Tips about graphs

Graphs from data: Where the axes and scales are not drawn, make sure that the scale occupies most or all of the given grid, is not an 'awkward' scale and that the plotted points occupy at least half of the height and width of the grid. Label the axes with the name, or symbol, of the variable with its unit – e.g. time / s, or F / N – and include scales. Plot points as accurately as possible; for points requiring interpolation between grid lines, the usual tolerance is $\pm \frac{1}{2}$ a square. Unless the question instructs differently, draw in the graph, don't just plot the points.

Sketch graphs: A sketch graph gives a good idea of the relationship between the two variables. It needs labelled axes but often it will not have scales and units. It is **not** an untidy ('hairy') graph. If the graph is intended to be a straight line, it should be drawn using a ruler. Sometimes significant values need to be labelled, e.g. include the amplitude, A, and the period, T, of an oscillation on a displacement–time graph.

Tips about calculations

If the command word is **calculate, find** or **determine**, full marks are given for the correct answer with no working shown. But an incorrect answer with no working scores 0. There are usually marks available for correct steps in the working even if the final answer is incorrect. Points the examiner will look for include:

- Selection of equation or equations and writing them down.
- Conversion of units, e.g. hours into seconds, mA into A.
- Insertion of values into equation(s) and manipulation of equation.
- Stating the answer – **remember the unit**.

If the command phrase is **show that**, the basic rules for the setting out are the same. You **must** give clear convincing steps: the examiner must be convinced that you know what you are doing and you will receive no marks for just stating the answer.

Tips about describing experiments

When describing one of the experiments from the specified practical work or for a practical which you are devising as part of the examination:

- Draw a simple diagram of the apparatus used **in its experimental arrangement**.
- Give a clear list of steps.
- Say what measurements are made and which instrument will be used.
- Say how the final determination will be made from your measurements.

Tips about questions involving units

This is mainly dealt with in Section 1.1 – Units and dimensions – of the AS SRG. One type of question requires you to suggest a unit for a quantity, which might not be on the specification. Such questions will always give an equation involving the quantity. The procedure to adopt is:

- Manipulate the equation to make the unknown quantity the subject.
- Insert the known units for the other quantities.
- Simplify.

Example

The drag coefficient, C_D, is defined by the equation $F_D = \frac{1}{2}\rho v^2 A C_D$, where F_D is the drag force on an object of cross-sectional area A moving at a speed v through a fluid of density ρ. Show that C_D has no units.

Answer

Make C_D the subject:
$$C_D = \frac{2F_D}{\rho v^2 A}$$

Rewrite in terms of units: The number 2 has no units:
$$\therefore [C_D] = \frac{[F_D]}{[\rho][v^2][A]} = \frac{\text{kg m s}^{-2}}{\text{kg m}^{-3}\,(\text{m s}^{-1})^2\,\text{m}^2} = \frac{\text{kg m s}^{-2}}{\text{kg m s}^{-2}}$$

The units of mass, length and time all cancel so C_D has no units. QED.

Note: Use the symbol [..] as shorthand for 'the unit of', e.g. $[v]$ = 'the unit of v'.

QER questions

Every examination paper will contain a question, or part of a question, which will test your ability to present a coherent account. These are called **Quality of Extended Response** questions and they are worth 6 marks. They could be AO1 questions on a piece of bookwork, e.g.

Explain how observations of the speed of objects in galaxies have been used as evidence for the existence of dark matter. [6 QER]

or a description of one of the pieces of specified practical work, e.g.

Describe in detail how you would make measurements on a gas to determine a value for absolute zero (0 K). [6 QER]

Whatever the topic of the question, the examiner will be looking for an identified set of ideas connected into 'a sustained line of reasoning which is coherent, relevant, substantiated and logically structured'. This means that the inclusion of incorrect or irrelevant material or poorly constructed arguments will be penalised. See Q&A 1 in Questions and answers section for an example of good and less good answers to a QER question.

Practice questions

Definition-type questions

1. A body undergoes *simple harmonic motion* with a *period* 2.5 ms and *amplitude* 16 cm. Say what is meant by each of the phrases in *italics*.

   [Vibrations – 1.6]

2. A body undergoes simple harmonic motion described by the equation:

   $$x = A \sin\frac{2\pi}{T}t$$

   Where x is the displacement, $A = 10$ cm, $T = 0.5$ s and t is the time.

   Sketch a graph of x against t between 0 and 2 seconds.

   [Vibrations – 1.6]

3. A system may undergo *free oscillations* or *forced oscillations*. Explain the difference between these two types of oscillation.

   [Vibrations – 1.6]

4. In an experiment to investigate resonance, a lightly damped oscillatory system of natural frequency f_0, is subject to a periodic driving force of constant amplitude and variable frequency f. The graph shows the variation of the amplitude, A, with f.

   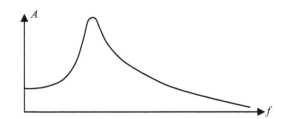

   (a) Label the graph with features of interest.

   (b) Add a second curve to show the expected behaviour with heavier but less than critical damping.

   [Vibrations – 1.6]

5. State the principle of conservation of momentum.

   [Dynamics – 1.3]

6. Explain what is meant by the Avogadro constant, N_A.

   [Kinetic theory – 1.7]

7. The first law of thermodynamics deals with energy transfers between a system and its surroundings. It may be written:

   $$\Delta U = Q - W$$

   Explain what is meant by each of the three terms in the equation.

   [Thermal Physics – 1.8]

8. The *specific heat capacity* of water is approximately $4200\,\text{J}\,\text{kg}^{-1}\,\text{K}^{-1}$.

    Define the term specific heat capacity and explain the above statement.

    [Thermal Physics – 1.8]

9. State Kepler's 3rd law of planetary motion and show how it can be derived for the case of circular orbits from Newton's law of gravitation.

    [Orbits and the wider universe – 2.8]

10. The diagram shows magnetic field lines linking a circuit at right angles.

    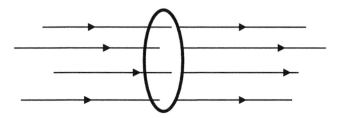

    Use the diagram to define the magnetic flux linking the circuit, defining any symbols you use.

    [Induction – 3.10]

11. The following equation relates to the decay of radioactive materials:

    $$A = \lambda N$$

    Define the symbols used in the equation and give their SI units.

    [Nuclear decay – 3.6]

Questions to test understanding

12. The momentum of a body is given by $p = mv$.

    The momentum of a photon is given by $p = \dfrac{h}{\lambda}$.

    Show that these two equations give the same units for p.

    [Dynamics – 1.3]

13. A planet is detected at a distance of $8 \times 10^{10}\,\text{m}$ from a star. It orbits the star at a constant speed of $5 \times 10^4\,\text{m}\,\text{s}^{-1}$.

    (a) Explain clearly how you can tell that the orbit is circular.

    (b) Find the following information about the planet and its orbit:

        (i) its angular speed,

        (ii) its orbital period,

        (iii) the frequency of the orbit

        (iv) the centripetal acceleration

    (c) Calculate the mass of the star

    [Orbits and the wider universe – 2.8]

14. The graph shows the resultant force, F, on a body of mass $2\,\text{kg}$ when displaced by x from its equilibrium position.

It is held at $x = 10.0\,\text{cm}$ and released at time $t = 0$.

Find its position and velocity at time $t = 1.5\,\text{s}$.

[Vibrations – 1.6]

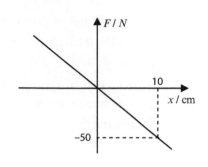

15. Newton's law of gravitation can be expressed in the following equation:

$$F = G\frac{M_1 M_2}{r^2}$$

(a) Identify the symbols used in the equation.

(b) The Moon orbits the Earth at a mean distance of $384\,000\,\text{km}$. The orbital period is 27.3 days. Use this information to obtain a value for the mass of the Earth. [You may assume that $M_{\text{Earth}} \gg M_{\text{Moon}}$.]

(c) The gravitational field strength at the surface of the Earth is $9.81\,\text{N}\,\text{kg}^{-1}$. The radius of the Earth is $6370\,\text{km}$. Use this information to obtain a second value for the mass of the Earth.

(d) Newton's law of gravitation refers to point masses. Explain how you were able to apply it when answering parts (b) and (c).

[Electrostatic and gravitational fields of force – 2.6]

16. The craters of the Moon have mainly been caused by the impact of orbiting bodies. This question concerns the energy released in such an impact.

A small asteroid, of diameter $50\,\text{m}$ and mean density $2500\,\text{kg}\,\text{m}^{-3}$, approaches the Earth–Moon system. Its speed at a large distance from the Earth–Moon is $1\,\text{km}\,\text{s}^{-1}$. It impacts on the Moon and causes a crater.

(a) Calculate the initial kinetic energy of the asteroid.

(b) Use the following data to calculate the gravitational potential at the surface of the Moon due to:

   (i) the gravitational field of the Moon;

   (ii) the gravitational field of the Earth.

     Data: Mass of Moon $= 7.35 \times 10^{22}\,\text{kg}$; radius of Moon $= 1740\,\text{km}$

     Mass of Earth $= 5.97 \times 10^{24}\,\text{kg}$; mean radius of Moon's orbit $= 384\,000\,\text{km}$

(c) Calculate the kinetic energy of the asteroid when it impacts the Moon.

[Electrostatic and gravitational fields of force – 2.6]

17. An electronic clock has a $0.2\,\text{F}$ backup capacitor in case its power supply is interrupted. The capacitor is initially charged to $3.3\,\text{V}$. The clock needs a minimum voltage of $1.3\,\text{V}$ to function.

When the supply is interrupted the capacitor starts discharging with a current of $1.0\,\mu\text{A}$. Estimate the number of hours before the clock stops working.

[Hint: Assume that the clock acts as a constant resistance load.]

[Capacitance – 2.4]

18. (a) The internal energy, U, of a capacitor can be calculated from the equation $U = \frac{1}{2}CV^2$. Starting from a definition of capacitance, show that the units of this equation balance.

(b) A 5.0 F capacitor is charged to 3.0 V and then isolated from the power supply.

   (i) Calculate the internal energy of the capacitor.

   (ii) A second, initially uncharged, 5.0 F capacitor is connected in parallel with the first capacitor. Calculate the total internal energy of the two capacitors.

   (iii) Comment on your answers to (i) and (ii).

[Capacitance – 2.4]

Data analysis questions

Note: These questions are longer than any data analysis question parts you can expect to meet in the component papers. All the question parts here, however, are examinable in these papers.

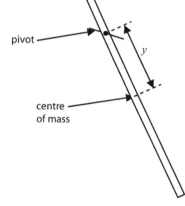

19. A group of Physics students investigated the oscillations of a 1.5 m long wooden beam. To do this they drilled a series of small holes at different distances, y, from the centre of mass. They suspended the beam from a nail at each of the holes in turn, released the beam from a small angle to the side measured the period, T, of oscillation using a stopwatch.

They read that T and y are related by the equation

$$T = 2\pi\sqrt{\frac{k^2 + y^2}{gy}}$$

where g is the acceleration due to gravity and k is a constant called the radius of gyration of the beam.

They repeated their measurements several times and obtained these data.

They estimated the uncertainty in T to be ±0.05 s and the uncertainty in y to be ±3 mm, because of the difficulty in estimating the positions of the centre of mass and the pivot.

y/m	T/s
0.700	1.98
0.600	1.89
0.500	1.90
0.400	1.87
0.300	1.93
0.200	2.15

(a) From the data, describe the relationship between T and y and relate it to the equation given above.

(b) Show that a graph of yT^2 against y^2 should be a straight line and state the relevance of the gradient and intercept to the quantities in the above equation.

(c) Complete the following table for each of the values of y and T given above.

y^2/m^2	$y\,T^2$/m s^2
0.490 ± 0.004	±
0.360 ± 0.004	±
0.250 ± 0.003	±
0.160 ± 0.002	±
0.090 ± 0.002	±
0.040 ± 0.001	±

(d) On the grid, plot the values of yT^2 against y^2. Plot the error bars in yT^2 [the error bars in y^2 can be omitted] and draw the steepest and least steep lines consistent with the data.

(e) Use your graph to determine values for k and g together with their absolute uncertainty.

20. For a certain class of thermistor, the resistance R varies with the kelvin temperature T according to the relationship:

$$R = Ae^{\frac{\varepsilon}{2kT}},$$

Where k is the Boltzmann constant, $1.38 \times 10^{-23}\,\mathrm{J\,K^{-1}}$, and ε is the band gap, which is the smallest energy difference between electrons which are bound to atoms and those which are free to move in the thermistor.

(a) Plan an experimental procedure to investigate the relationship between R and T between 273 K (0°C) and 373 K.

(b) The following results were obtained.

T/K	R/Ω
273	380
298	100
323	38.5
348	14.7
373	6.2

Plot a suitable graph to test the relationship.

(c) Comment on whether the results support the suggested relationship.

(d) Use your graph to determine ε.

Questions and answers

This part of the guide looks at actual student answers to questions. There is a selection of questions covering a wide variety of areas of study. In each case there are two answers given; one from a student (Seren) who achieved a high grade and one from a student who achieved a lower grade (Tom). We suggest that you compare the answers of the two candidates carefully; make sure you understand why one answer is better than the other. In this way you will improve your approach to answering questions. Examination scripts are graded on the performance of the candidate across the whole paper and not on individual questions; examiners see many examples of good answers in otherwise low scoring scripts. The moral of this is that good examination technique can boost the grades of candidates at all levels.

Q&A 1

The rotation curve for a spiral galaxy is shown

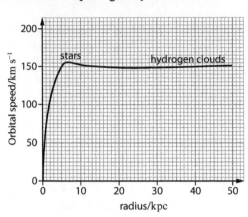

Explain how such graphs are obtained and how their shape informs our theories of the make-up of the universe.

[6 QER]

Tom's answer

Astronomers look at stars and hydrogen clouds and measure their red-shifts and blue-shifts, which tell us their speed. The strange thing they notice is that the objects in the galaxy are moving so fast that they should escape – the galaxy shouldn't hold together. But it does so because of dark matter. There is about five times more dark matter than normal matter. The calculated rotation curve looks like this:

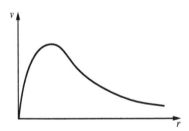

Examiner commentary

Tom has covered both aspects of the question and he presents some nuggets of knowledge but they are not linked well together into a coherent account. The examiner is not going to ignore the graph (in spite of the QER rule that the answer should be written) but it is not put into any context so it is of limited use. Much detail is missing, such as how Doppler shifts are used.

Conclusion

The lack of structure makes this a lower band answer. An examiner would probably give him 2 marks out of 6.

Seren's answer

We measure orbital speeds in galaxies using the Doppler effect on the wavelength of lines in the spectrums of stars and hydrogen clouds. These wavelengths are measured, compared with the laboratory value and the radial velocity calculated using the equation $\frac{\Delta\lambda}{\lambda} = \frac{v}{c}$.

For the hydrogen clouds, the rotation speeds are much higher than we would expect from the estimated mass of the normal matter in the galaxy. These clouds are outside the visible galaxy where there is hardly any material, so we would expect the orbital speed to decrease with radius according to $v = \sqrt{\frac{GM}{r}}$. However, the velocity is nearly constant. These two things mean that: (a) there is a lot more mass than we can see in the galaxy and (b) it extends out to these gas clouds. We think this missing mass is provided by dark matter – the visible galaxy is embedded in it.

Examiner commentary

This is a top-band answer. Seren covers both parts of the question. There are missing elements: for example, she uses equations and doesn't define the terms (but she could have done without the second equation); also she doesn't say that in the absence of dark matter, the orbiting clouds would escape from the gravity of the galaxy. Some comment on the nature of dark matter would also have been appropriate. However, the detail given in the previous sentences balances this.

Conclusion

A generous examiner would award her 6 marks out of 6. A less generous examiner would award 5 marks.

A gas undergoes the cycle ABCDA as shown in the p–V graph below.

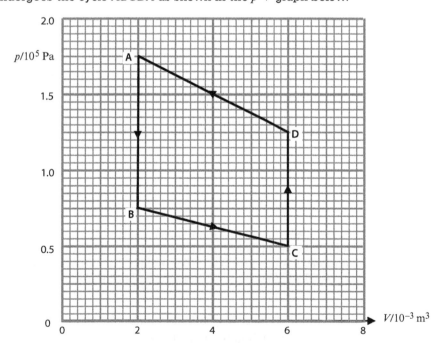

(a) Explain very briefly why no work is done during AB or CD. [1]

(b) Calculate the work done by the gas during process DA. [3]

(c) The first law of thermodynamics is usually written $\Delta U = Q - W$. State the meaning of each term. [ΔU, Q and W] [3]

(d) Calculate the heat flow out of the gas during the cycle ABCDA. [3]

Tom's answer

(a) There is no change in volume. ✓

(b) Work done $= p\Delta V = \frac{1}{2}bh + bh = \frac{1}{2}(4 \times 0.5) + 4 \times 1.25$ ✓

$\qquad = 6\,\text{J}$

(c) ΔU: The change in internal energy ✓
$\qquad Q$:　Internal heat flow ? ✗
$\qquad W$:　Work done ✗

(d) D→A: 6 J
$\qquad$ B→C: 2.5 J
$\qquad$ Total heat $= 3.5\,\text{J}$ ✓ [just]

Seren's answer

(a) Because the volume remains constant. ✓ There is no volume change so work done $= 0$ since $W = p\Delta V$.

(b) Work done = area under graph =

$$\frac{1.75 \times 10^5 + 1.25 \times 10^5}{2} \times 4 \times 10^{-3}\ \checkmark = 600\,\text{J} \checkmark$$

Work is negative because ΔV is negative, so work done by gas $= -600\,\text{J}$ ✓

(c) ΔU: The change in internal energy ✓
$\qquad Q$:　The heat flow into the system ✓
$\qquad W$:　The work done by the system ✓

(d) $\Delta U = 0$, $\therefore Q = W$ ✓
$\qquad W = W_{DA} - W_{BC} = -600 + \frac{1}{2} \times (0.755 \times 10^5 + 0.50 \times 10^5)$
$\qquad\qquad \times 4 \times 10^{-3} = -350\,\text{J}$
$\qquad \therefore Q = -350\,\text{J}\checkmark$, so heat flow out of the gas $=$
$\qquad + 350\,\text{J}$ ✓

Examiner commentary

(a) Tom hits the marking point.

(b) Tom has made a good attempt to calculate the area so has obtained the first mark, but he has failed to spot the multiplying factors on the scales of the graph. He also fails to understand that the work is done **on** the gas and so he needs to introduce a minus sign.

(c) The examiner hesitated by 'internal heat flow' but Tom hasn't clearly stated that Q is the heat flowing into the system. Similarly Tom needs to give a direction to W.

(d) Tom appears to be calculating work. He notices that the two quantities of energy flow are in different directions and the difference needs to be found, for which he gains the mark. He does not relate Q to W. He would not be penalised twice for the multiplying factors [see part (b)].

Tom scores 4 out of 10 marks.

Examiner commentary

(a) On this occasion, the examiner was only looking for the statement about volume, hence the instruction to explain 'very briefly'. Seren gave more details which is a good idea.

(b) The first mark is for an attempt to calculate the area – in this case it was successful! The final mark is for the negative sign.

(c) All correct. '. . . of the system' would have been good to see in the explanation of ΔU, but was not required.

(d) The statement that $Q = W$ is an important part of the answer.

Again, the sign of the answer is important. Because net work is done **on** the gas in taking it round the cycle, heat must flow **out**.

Seren scores 10 out of 10 marks.

Q&A 3

Describe an everyday circumstance where resonance occurs. Your example of resonance may be useful or it may be an example where resonance should be avoided. You should explain what your oscillating system is, what provides the driving force and the result of the resonance. A diagram may (or may not) assist your answer. [4]

Marking scheme

Diagram/statement of application [e.g. bridge, car rattle...] ✓
Description of plausible oscillating driving force ✓
Description of system ✓
Large amplitude because of same frequency [or graph showing resonance, with labelled axes] ✓

Tom's answer

Everyday example of resonance can be that in the bridges ✓ *where driving force is equal to the natural frequency (Millennium Bridge). People's footsteps, when they walk in time, provide the driving force* ✓. *The bridge will collapse as it disturbs.*

Examiner commentary

Tom has identified that the situation involves a bridge and has gained the first mark. He has also indicated that it is the walking people that provide driving force. He has not indicated clearly what the oscillating system is, or the conditions under which resonance occurs. The statement about the bridge collapsing is too vague to achieve the 'effect' mark.

A better answer, in the context of the Millennium Bridge, would have included: the effect of the footsteps was to make the bridge oscillate from side to side; because the frequency of the footsteps coincided with the natural frequency of the bridge, large amplitude oscillations occurred.

Tom scores 2 out of 4 marks.

Seren's answer

Resonance occurs when a child is being pushed on a swing ✓ *by an adult. The adult provides the driving force* ✓ *and the swing oscillates about a central equilibrium position.* ✓ *If the adult provides a force [pushes!] with the same frequency as the natural frequency of the swing, large amplitude oscillations will occur, i.e. the swing will swing high!*✓

Examiner commentary

Seren's answer is not perfect. The second point should perhaps have stated 'The regular pushes of the adult...'. The third point, which required identification of the oscillating system, is a little incomplete too: better would be 'the child, the swing seat and the supporting wires form the oscillating system, which swings to and fro'. However, the examiner considered she had done enough for the marks!

Seren scores 4 out of 4 marks.

The diagram shows an isolated negative charge ($-Q$).

$-Q$

(a) Sketch electric field lines and equipotential surfaces for the negative charge. [3]

(b) Two point charges are placed a distance 8.0 cm apart as shown in the diagram below:

8.0 cm

+2.0 μC   +2.0 μC

   (i) Calculate the force between the two charges. [2]
   (ii) One of the two +2.0 μC charges is released from rest while the other is held stationary. Use the concept of potential or potential energy to calculate the maximum kinetic energy that the charge will eventually acquire. [3]

(c) Calculate the resultant electric field at P in the set-up shown below: [4]

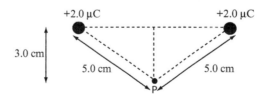

+2.0 μC   +2.0 μC

3.0 cm

5.0 cm   5.0 cm

P

Tom's answer

(a)

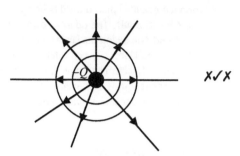

$$\times\checkmark\times$$

(b) (i) $F = \dfrac{1}{4\pi\varepsilon_0}\dfrac{2\times10^6 \times 2\times10^6}{(8\times10^{-2})} = 5\times10^{-11}\times\dfrac{1}{4\pi\varepsilon_0}$ ✗

(ii) $V = \dfrac{1}{4\pi\varepsilon_0}\dfrac{Q}{r} = 2.25\times10^5$

∴ Max kinetic energy $= 2.25\times10^5$ J ✓ e.c.f

(c)

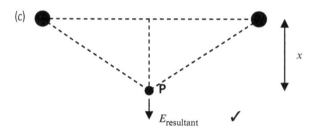

$E_{\text{resultant}}$ ✓

$E_{\text{at P}} = \dfrac{1}{4\pi\varepsilon_0}\dfrac{Q}{r^2} = 7.19\times10^8$ ✓✗ $x = 5.75\times10^8$

∴ resultant at $P = 2\times5.75\times10^8$
$= 1.19\times10^9$ C ✗

Seren's answer

(a)

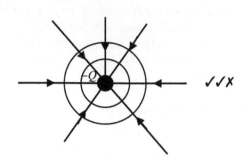

$$\checkmark\checkmark\times$$

(b) (i) $F = \dfrac{Q_1Q_2}{4\pi\varepsilon_0 r^2} = \dfrac{2\times10^{-6}\times2\times10^{-6}\times9\times10^9}{(8\times10^{-2})^2}$ ✓

$= 5.6$ N C^{-1} ✓ [unit not penalised here]

(ii) $V = \dfrac{Q_1Q_2}{4\pi\varepsilon_0 r} = \dfrac{2\times10^{-6}\times2\times10^{-6}\times9\times10^9}{8\times10^{-2}}$ ✓

$= 0.45$ J. ✓

PE lost $=$ KE gained ✓

(c)

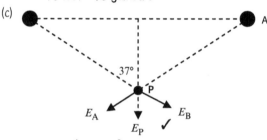

$E_A = \dfrac{2\times10^{-6}\times9\times10^9}{(5\times10^{-2})^2} = 7\,200\,000$ ✓✓

$E_B = 7\,200\,000$

$E_P = E_A \cos37° + E_B \cos37° = 11\,520\,000$ N C^{-1}

Examiner commentary

(a) Tom has drawn field lines but they are in the wrong direction. He has drawn two equipotentials. He has not identified the field lines and equipotentials.

(b) (i) Tom has used an incorrect formula – the 8×10^{-2} should be squared – so no marks are given.

(ii) Tom has calculated a potential and thinks this is the potential <u>energy</u>. He gains a mark for realising that the drop in potential energy is the gain in kinetic, even though the figure is wrong.

(c) Tom realises that the resultant field at P is vertically downwards – 1 mark. His second mark is for the use of the formula for the electric field at P due to one of the $2.0\,\mu$F charges – his answer is wrong as is the use of the x distance subsequently.

Tom scores 4 out of 12 marks.

Examiner commentary

(a) The missing mark is because Seren has not identified the equipotentials and the field lines.

(b) (i) Seren has used the correct formula. She has used the very good approximation, that $(4\pi\varepsilon_0)^{-1} = 9\times10^9$ F^{-1}m. .

(ii) Seren has made a slip in writing '$V =$' rather than 'Potential energy $=$' at the start of her answer. She goes on to treat the answer as PE, so it is accepted.

(c) Seren realises that the resultant field at P is vertically downwards – 1 mark. She correctly calculates the contribution to E_P from A and B [her labels] but unfortunately has miscalculated the angle at P [see diagram] – it is 53°.

Seren scores 10 out of 12 marks

The product of the pressure and volume of an ideal gas may be expressed as

$$pV = nRT.$$

The product may also be written in terms of the mean square speed of the molecules as

$$pV = \tfrac{1}{3}Nm\overline{c^2}$$

(a) Derive in clear steps a formula that shows how the internal energy of the ideal gas depends on the temperature of the gas. [4]

(b) A canister of volume 0.025 m³ contains helium gas at a pressure of 305 kPa and a temperature of 18°C. Calculate:

(i) the internal energy of the gas [2]

(ii) the number of molecules of helium in the canister. [2]

Tom's answer

(a) Combining $pV = nRT$ and $pV = \tfrac{1}{3}Nm\overline{c^2}$ we get

$nRT = \tfrac{1}{3}m\overline{c^2}$ ✓

The energy of the gas is the kinetic energy of the molecules ✓ $= \tfrac{1}{2}m\overline{c^2}$ ✗

So the internal energy is $\tfrac{3}{2}nRT$ ✗

(b) (i) From part (ii), the number of moles is 0.0510

So $U = \tfrac{3}{2}nRT = \tfrac{3}{2} \times 0.0510 \times 8.31 \times 18 = 11.4\,\text{J}$ ✗✓ **ecf**

(ii) $n = \dfrac{pV}{RT} = \dfrac{305 \times 0.025}{8.31 \times 18} = 0.0510$ ✗

∴ Number of molecules $= 0.0510 \times 6.02 \times 10^{23} = 3.07 \times 10^{22}$ ✓ **ecf**

Examiner commentary

(a) Tom starts well by combining the two given equations and identifying the internal energy of the gas with the molecular kinetic energy, which he unfortunately gives as the mean energy of an individual molecule. He then mysteriously writes the equation which is given in the Data booklet.

(b) Tom answers this in an unusual way by first answering part (ii) and using it to answer part (i). This is legitimate and the examiner has applied the e.c.f. rules accordingly, though the mark allocation for part (i) becomes rather generous.

(ii) Tom uses the correct equation for calculating the number of moles but unfortunately he has fallen into both unit traps: he should have converted to Pa and K. However, he then goes on to use n to calculate the number of molecules and gains the second mark.

(i) The ecf value of n is accepted but he again loses out on the unit trap.

Tom scores 4 out of 8 marks.

Seren's answer

(a) Ideal gas molecules don't have forces between them so the internal energy is just their kinetic energy. ✓

∴ Internal energy, $U = \tfrac{1}{2}Nm\overline{c^2}$ ✓

From equation 2, $Nm\overline{c^2} = 3pV$, so $U = \tfrac{1}{2} \times 3pV = \tfrac{3}{2}pV$ ✓

So, from equation 1, $U = \tfrac{3}{2}nRT$ ✓

(b) (i) $U = \tfrac{3}{2}pV = \tfrac{3}{2} \times 305 \times 10^3\,\text{Pa} \times 0.025\,\text{m}^3$ ✓

$= 11\,4000\,\text{J (3 s.f.)}$ ✓

(ii) $n = \dfrac{pV}{RT} = \dfrac{305 \times 10^3\,\text{Pa} \times 0.025\,\text{m}^3}{8.31\,\text{J K}^{-1} \times 291\,\text{K}}$

$= 3.15\,\text{mol}$ ✓

Examiner commentary

(a) A good answer by Seren. She makes the important point about the lack of intermolecular forces and correctly states the internal energy as $U = \tfrac{1}{2}Nm\overline{c^2}$. Following this she logically and clearly uses the given equation to derive the Data booklet formula. Note that the mark is for the working towards the equation and not for the equation itself.

(b) (i) Seren has seen the easiest way of tackling this using the $U = \tfrac{3}{2}pV$, which is not in the Data booklet, but arose from her part (a) working. She skilfully avoids the unit trap.

(ii) Again, Seren converts both units and correctly calculates the number of moles of the gas. Unfortunately she fails (forgets?) to go on to calculate the number of molecules.

Seren scores 7 out of 8 marks.

Q&A 6

(a) Calculate the binding energy per nucleon of $^{14}_{6}C$. [4]

[$1u \equiv 931$ MeV, $m_{neutron} = 1.008665$ u, $m_{proton} = 1.007276$ u, mass of $^{14}_{6}C$ nucleus = 13.999950 u]

The following reaction can be regarded as evidence for the existence of neutrinos (or an anti-neutrino in this case).

$$^{14}_{6}C \longrightarrow {}^{14}_{7}N + {}^{0}_{-1}\beta + \bar{\nu}_e$$

Mass of $^{14}_{6}C$ nucleus = 13.999950u; Mass of $^{14}_{7}N$ nucleus = 13.999234u
Mass of β^- particle = 0.000549u; The mass of the anti-neutrino $\bar{\nu}_e$ is negligible.

(b) Calculate the energy released in this reaction. [3]

The evidence for the existence of the anti-neutrino came from the (unexpected) wide variation of the energies of the β-particles emitted. However, you should now ignore the existence of the anti-neutrino.

(c) Explain briefly, using conservation of momentum, which particle (N or β⁻) receives most of the energy of the reaction. [3]

$^{14}_{6}C$

Before the reaction (stationary $^{14}_{6}C$)

$^{14}_{7}N$

After the reaction

Tom's answer

(a) $m_{total} = 6 \times 1.007276 + 8 \times 1.008665 ✓ = 14.112976$ u
$m_{lost} = 14.112976 - 13.999950 = 0.113026$ u ✓
$E = mc^2 = 0.113026 \times (3 \times 10^8)^2 = 1.017234 \times 10^{16}$ eV ✗

(b) $13.999950 - (13.999234 + 0.000549) = 1.67 \times 10^{-4}$u ✓
$E = mc^2 = 1.67 \times 10^{-4} \times (3 \times 10^8)^2 = 1.503 \times 10^{13}$ eV

(c) The nitrogen nucleus receives most of the energy from the reaction because it is a lot heavier than the β / electron emitted ✓ and the $^{14}_{7}N$ still has a fair bit of velocity. ∴ the $^{14}_{7}N$ has the most energy from the collision.

Examiner commentary

(a) Tom has calculated the mass of the 14 nucleons correctly and the mass deficit of the nucleus in u – giving him the first two marks. He wants to use $E = mc^2$, so he should have converted the u to kg. He also fails to divide by the number of nucleons.

(b) Tom's first step is correct but he makes the same mistake as in part (a).

(c) Tom correctly identifies the significant fact of the relative masses of the $^{14}_{7}N$ and the β-particle but fails to draw the correct conclusion.

Tom scores 4 out of 10 marks.

Seren's answer

(a) $6 \times 1.007276 + 8 \times 1.008665 ✓ = 14.112976$ u
$14.112976 - 13.999950 = 0.113026$ u ✓
$\dfrac{0.113026 \times 931}{14} ✓ = 7.52$ MeV ✓

(b) $13.999950 - (13.999234 + 0.000549) = 1.67 \times 10^{-4}$u ✓
Energy released = $931 \times 1.67 \times 10^{-4} ✓ = 0.155$ MeV ✓

(c) The momentums of the nucleus and the β particle, ie mv for the two particles are the same and so v for the lighter β particle ✓ is much higher ✓. Kinetic energy is $\frac{1}{2}mv^2$ which is $\frac{1}{2}mv \times v$, so the β particle with the bigger velocity has the bigger KE ✓.

Examiner commentary

(a) Seren has done all the steps and has received the marks. Her communication is not perfect because she doesn't say what each line of the answer means. Her answer is correct [the units MeV, MeV/nucleon were accepted].

(b) Again correct with better communication.

(c) Seren has come to the correct conclusion with good reasoning. Even easier would be to note that the relationship between KE and momentum is KE = $\dfrac{p^2}{2m}$, so if the momenta are the same, the lighter particle gets the lion's share of the energy.

Seren scores 10 out of 10 marks.

Caesium-137 is a radioactive by-product from fission nuclear power stations. It has a half-life of 30 years and emits β⁻ radiation.

(a) Complete the following reaction equation: [2]

$$^{137}_{55}\text{Cs} \longrightarrow \text{.......Ba} + \text{.......}\beta^-$$

(b) Show that the decay constant of caesium-137 is approximately 7×10^{-10} s⁻¹. [2]

(c) Show that the initial activity of 1.0 kg of caesium-137 is approximately 3×10^{15} Bq. [2]

(d) Explain why 1.0 kg of caesium-137, although it has an activity of 3×10^{15} Bq, would be quite safe in a sealed metal box of thickness 1 cm. [1]

(e) When the activity of 1.0 kg of caesium has dropped to 1000 Bq (comparable to soil) it can be disposed of by mixing with soil and scattering on the ground. Calculate how long it takes for the caesium sample to reduce its activity from 3×10^{15} Bq to 1000 Bq. [3]

Tom's answer

(a) $^{137}_{55}\text{C} \longrightarrow {}^{137}_{56}\text{Ba} + {}^{0}_{0}\beta^-$ ✗✗

(b) $\lambda = \dfrac{\ln 2}{T_{\frac{1}{2}}} \checkmark = \dfrac{0.693}{30 \times 365 \times 24 \times 60 \times 60} = 2.198 \times 10^{-8}$ s ✗

(c) $A = \lambda N \checkmark = 3 \times 10^{15}$ Bq

(d) The beta particles would be absorbed by the metal. ✓

(e) $A = A_0 e^{-\lambda t}$

$\ln A = \ln A_0 - \lambda t \checkmark$

$\lambda t = \ln A_0 - \ln A$

$t = \dfrac{\ln\left(\dfrac{A_0}{A}\right)}{\lambda} = \dfrac{\ln\left(\dfrac{3 \times 10^{15}}{1000}\right)}{2.198 \times 10^{-8}} \checkmark = 567\,657\,927$ s $= 18$ years.

Examiner commentary

(a) Tom misses both marks. The first was for applying the conservation of A and Z and the second was for all the numbers being correct.

(b) The incorrect answer was down to a calculator slip. The mark was actually for the correct expression, which Tom gave – unfortunately his answer contradicted this expression (he forgot to divide by 30).

(c) The first mark was for the selection of the correct equation, which earned the mark. In a 'show that' question, all steps in the answer must be given.

(d) Correct.

(e) Another calculator slip. One mark was for taking logs correctly; the second for the correct identification of A and A_0.

Tom scores 5 out of 10 marks.

Seren's answer

(a) $^{137}_{55}\text{C} \longrightarrow {}^{137}_{56}\text{Ba} + {}^{0}_{-1}\beta^-$ ✓✓

(b) $\lambda = \dfrac{\ln 2}{T_{\frac{1}{2}}} \checkmark = \dfrac{0.693}{30 \times 365 \times 24 \times 60 \times 60} \checkmark = 7.33 \times 10^{-10}$ s⁻¹

(c) $A = \lambda N \checkmark$ $N = \dfrac{1000}{137} \times 6.02 \times 10^{23} = 4.394 \times 10^{24}$.

So $A = 7 \times 10^{-10} \times 4.394 \times 10^{24} = 3.0758 \times 10^{15}$ Bq ✓

(d) It only emits β radiation

(e) $A = A_0 e^{-\lambda t}$ ✓ $1000 = 3 \times 10^{15} e^{-\lambda t}$

$1000 = 3 \times 10^{15} \times e^{-7 \times 10^{-10} t}$

so $\ln 1000 = \ln 3 \times 10^{15} - 7 \times 10^{-10} t$ ✓

So $t = \dfrac{\ln 3 \times 10^{15} - \ln 1000}{7 \times 10^{-10}} = 4.104 \times 10^{10}$ s ✓

Examiner commentary

(a) The totals of A and Z on the two sides are equal and the figures are all correct.

(b) Correct working.

(c) Seren's working is correct. She has used the approximate value for λ, which is acceptable. In this 'show that' type of question it is best to give the asked-for figure to at least one more s.f. than requested.

(d) Seren should have related the question to the penetrating powers of β-particles.

(e) Again, Seren has used the given approximate values for λ and A_0.

Seren scores 9 out of 10 marks.

Q&A 8

Electrons move through a metallic conductor as shown and experience a force due to the applied magnetic field (B perpendicular to the front face as shown).

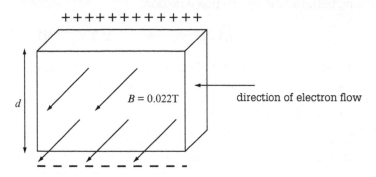

$B = 0.022\text{T}$

direction of electron flow

(a) Explain why charges accumulate on the upper and lower face of the conductor as shown. [2]

(b) Indicate on the diagram how you would connect a voltmeter in order to measure the Hall voltage (V_H). [1]

(c) By equating the electrical and magnetic forces acting on an electron in the conductor, show that $V_H = Bvd$. [3]

(d) (i) The magnetic field ($B = 0.022\,\text{T}$) is produced by a solenoid of length $2.00\,\text{m}$ and with 15000 turns. Calculate the current in the solenoid. [2]

    (ii) Where must the conductor be placed and how should it be orientated in relation to the solenoid to obtain the maximum Hall voltage? [2]

Tom's answer

(a) The force produced by the electrons moving through the conductor and the B-field generated push the electrons down to the lower face. ✓

(b)

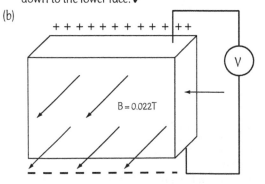

(c) $F = Bqv$ $F = Ee$ $Bqv = Ee$ ✓

(d) (i) $B = \mu_0 n \ell$

    $\rightarrow I = \dfrac{B}{\mu_0 n}$ ✓ $= 1.167\,\text{A}$ ✗

   (ii) It should be parallel to the solenoid ✗ and placed inside the coil ✓

Examiner commentary

(a) Tom has communicated that the electrons experience a downward force and thus gains one mark. He needs to give more detail, e.g. correctly using Fleming's LH rule to gain the second.

(b) Correct – voltmeter connected between the upper and lower surfaces.

(c) A good start, Tom equates the electric and magnetic forces but makes no further progress.

(d) (i) The incorrect use of ℓ rather than I is ignored as I is used in the manipulated form of the equation. Tom's mistake is not to realise that n is the number of turns <u>per metre</u>. He therefore needs to divide n by 2, so the value for I should be double his answer.

   (ii) The plane of the conductor should be perpendicular to the field and hence to the axis of the solenoid.

Tom scores 5 out of 10 marks.

Seren's answer

(a) The electrons feel a force downwards ✓ due to Flemming's left-hand rule✓. This causes the top to become positive and the bottom to become negative.

(b)

$+ + + + + + + + + + + +$

$B = 0.022T$

V

$- - - - - - - - - -$

(c) $Ee = Bev$ ✓

$E = \dfrac{V}{d}$, so $\dfrac{V_H}{d}e = Bev$ ✓, $\therefore V_H = \dfrac{Bevd}{e}$ ✓ $= Bvd$

(d) (i) $B = \mu_0 nI$.

$I = \dfrac{B}{\mu_0 n} = \dfrac{0.022}{\mu_0 \times 7500} = 2.33\ A$ ✓✓

(ii) The conductor should be placed perpendicular to the B-field to obtain a max Hall voltage – perpendicular ✓ (b.o.d.) inside the solenoid ✓.

Examiner commentary

(a) The misspelling of Fleming is ignored.

(b) Seren has placed the voltmeter correctly.

(c) A good clear derivation.

(d) (i) Seren has remembered to divide the number of turns by the length to find n.

(ii) Only just. It is not very clear the large flat surfaces of the conductor should be placed perpendicular to <u>the axis of the solenoid</u> – the examiner had to infer this – but she has been given the benefit of the doubt on this occasion.

Seren scores 10 out of 10 marks.

Q&A 9

A magician's metallic wand can spring apart into the shape of a circular hoop (see below).

 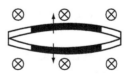 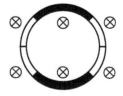

(a) The hoop is in a magnetic field. Explain why an EMF is induced in the hoop as it expands. [3]

(b) Explain why the current flows anticlockwise in the diagram. [2]

(c) The hoop, of radius 31.0 cm, is in a region of uniform magnetic flux density (B) of 58 mT and expands from the wand shape to the hoop in a time of 63 ms. Calculate the mean current flowing in the hoop as it expands if the resistance of the hoop is 0.44 Ω. [5]

Tom's answer

(a) The sides of the hoop move through ✗ magnetic field lines ✓ as the hoop expands. This induces a current.

(b) The current flows anticlockwise because the B-field is directed at a right angle to the hoop and is flowing towards the top of the hoop. ✗

(c) $Area = \pi r^2 = \pi \times 0.31^2$ ✓

$\Phi = BA$ ✓ $= 58 \times 10^{-3} \times \pi \times 0.31^2 = 1.75 \times 10^{-2}$

Change of flux per second $= \dfrac{1.75 \times 10^{-2}}{63 \times 10^{-3}} = 0.2779$ ✓

$I = VR$ ✗ $= 0.2779 \times 0.44 = 0.122$ A

Examiner commentary

(a) Tom has not mentioned that the hoop sides <u>cut</u> the field lines. Moving along field lines produces no current. For the third mark a correct mention of a law of induction or a relevant equation is needed.

(b) Tom's answer lacks any substance. He could have considered the top of the hoop and applied the right-hand rule to show that the induced current is to the left, which is anticlockwise.

(c) Tom has correctly worked out the area of the expanded loop and the flux which links it. Unaccountably he has made a mistake with the well-known equation $V = IR$ and so failed to calculate the current correctly.

Tom scores 4 out of 10 marks.

Seren's answer

(a) EMF is the rate of change of flux linkage✓. As the area of the wand increases✓, the flux linking the loop increases✓, so inducing an EMF.

(b) The current must oppose the change, e.g. on the top of the loop the force must be downwards, so the current must be to the left by the left-hand rule. ✓✗

(c) $V_{induced} = B\dfrac{dA}{dt}$ ✓ $= 58 \times 10^{-3} \times \dfrac{\pi \times 0.31^2}{63 \times 10^{-3}}$ ✓✓ $= 0.278$ V

$I = \dfrac{V}{R} = \dfrac{0.278 V}{0.44\Omega}$ ✓ $= 0.632$ A ✓

Examiner commentary

(a) Seren has used a different approach from Tom. Both are valid. Seren has mentioned the general point relating EMF to change of flux linkage and pointed out how these apply to the hoop.

(b) If Seren had said that the current in the bottom of the loop was to the right she'd have received the second mark.

(c) Seren's answer uses calculus notation correctly. This is not required. Tom's approach would have gained all 5 marks if he had used the correct equation for his final step.

Seren scores 9 out of 10 marks.

(a) Calculate the capacitance of the capacitor shown. [2]

(b) The capacitor is charged so that there is a pd of 1.2 kV across the plates. Calculate:
   (i) the charge stored, [1]
   (ii) the energy stored in the capacitor. [1]

(c) The capacitor is discharged through a 670 kΩ resistor. Calculate the time the capacitor takes to lose half its charge. [3]

(d) Explain briefly whether or not the time the capacitor takes to lose half its energy is longer or shorter than your answer to (c). [2]

(e) An electron is located between the plates of the charged capacitor. Show that the acceleration experienced by the electron is approximately 6×10^{17} m s^{-2}. [3]

(f) The electron starts from rest halfway between the plates.
   (i) Calculate the speed of the electron when it strikes the upper plate of the capacitor. [2]
   (ii) Show that the speed of the electron (when it strikes the upper plate of the capacitor) corresponds to a kinetic energy of 0.6 keV and explain briefly another method for obtaining this answer of K.E. = 0.6 keV. [3]
   (iii) Calculate the time the electron takes to travel to the upper plate. [3]

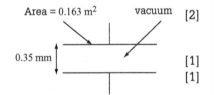

Area = 0.163 m², vacuum

0.35 mm

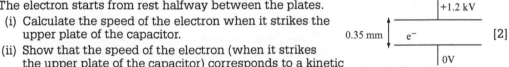

+1.2 kV

0.35 mm, e⁻

0V

Tom's answer

(a) $C = \dfrac{\varepsilon_0 A}{d}$

$C = \dfrac{8.85 \times 10^{-12} \times 0.163}{0.33 \times 10^{-3}}$ ✓ (only) $= 4.37 \times 10^{-9}$

(b) (i) $Q = CV = 4.37 \times 10^{-9} \times 1.2 \times 10^3$

$\qquad = 5.244 \times 10^{-6}\,C$ ✓ e.c.f.

(ii) $E = \frac{1}{2} QV$

$\qquad E = 5.244 \times 10^{-6} \times 1.2 \times 10^3 = 6.293 \times 10^{-3}\,J$ ✗

(c) $Q = Q_0 e^{-t/RC}$. $RC = 670 \times 10^3 \times 4.37 \times 10^{-9} = 2.93 \times 10^{-3}\,s$ ✓

$\ln Q = \ln Q_0 - \dfrac{t}{RC}$ $\qquad \dfrac{\ln Q}{\ln Q_0} = -\dfrac{t}{RC}$ ✗

$t = \dfrac{\ln 2.62 \times 10^{-6} \times 2.93 \times 10^{-3}}{\ln 5.24 \times 10^{-6}} = 3.10 \times 10^{-3}\,s$

(d) Shorter, as energy is released to the outside, extra escapes ? ✗

(e) $F = \dfrac{1}{4\pi\varepsilon_0} \dfrac{1.2 \times 10^3 \times 1.6 \times 10^{-19}}{(0.35 \times 10^{-3})^2} = 14.1\,N$ ✗

$a = \dfrac{14.1}{9.11 \times 10^{-31}} =$

(f) (i) $v^2 = u^2 + 2ax$ ✓

$\qquad v^2 = 2 \times 6 \times 10^{17} \times 0.00035\,m = 4.2 \times 10^{14}$ ✗

$\qquad v = 2.05 \times 10^7\,m\,s^{-1}$

(ii) $ke = \frac{1}{2} m v^2 = \frac{1}{2} \times 9.11 \times 10^{-31} \times (2.05 \times 10^7)^2$ ✓

(iii) $v = u + at; t = \dfrac{v}{a} = \dfrac{2.05 \times 10^{-7}}{6 \times 10^{17}}$ ✓✓ ecf

$\qquad t = 3.42 \times 10^{-11}\,s$ ✓

Examiner commentary

(a) Tom started well and obtained the mark for using the capacitor equation. Unfortunately he made a slip, used an incorrect values for d, and paid for it. He lived dangerously by omitting the unit!

(b) (i) Good – note that Tom used his (incorrect) value for C but was credited on the error carried forward principle.

(ii) Another slip – this time Tom omitted the factor of $\frac{1}{2}$ in his calculation.

(c) Tom correctly calculated the value of the time constant, RC, and received a mark. He incorrectly manipulated the equation after correctly taking logs. He might have done better to simplify the first equation – using Q and $\frac{1}{2}Q$ and cancelling the Qs – to start with. It would have produced an easier equation to work with.

(d) The answer 'shorter' is correct but it needs to arise from correct reasoning.

(e) Tom appears to be attempting to apply the equation for the force between two point charges, which is inappropriate. No credit can be given.

(f) (i) Correct equation – on this occasion, this was enough for the first mark.

His value of x was incorrect – it should have been $\frac{1}{2} \times 0.35$ mm $= 0.175$ mm.

(ii) Tom could have earned a second mark by converting his value of energy to eV. It would have been incorrect but ecf would have come to his rescue!

(iii) A good final answer. It is worth persevering to the end of a question – the last part is not always the most difficult and Tom snapped up 3 marks.

Tom scores 8 out of 20 marks.

Seren's answer

(a) $C = \dfrac{\varepsilon_0 A}{d} = 4.1216 \times 10^{-9} \, F$ ✓✓

(b) (i) $\quad Q = CV = 4.1216 \times 10^{-9} \times 1.2 \times 1000 = 4.946 \times 10^{-6} \, C$ ✓(b.o.d.)

(ii) $\quad E = 0.5QV$

$\quad\quad = 0.5 \times 4.946 \times 10^{-6} \times 1.2 \times 1000 = 2.968 \times 10^{-7} \, J$ ✗

(c) $Q = Q_0 e^{-t/RC} \quad\quad R = 670$

$\dfrac{Q_0}{2} = Q_0 e^{-t/RC}$ ✓ $\quad \dfrac{-t}{RC} = \ln\dfrac{1}{2}$

$\quad t = -\ln\dfrac{1}{2} \times 670 \times 1000 \times 4.1216 \times 10^{-9} = 1.91 \times 10^{-3} \, s$ ✓✓

(d) The rate of charge loss is exponential. The energy is the product of charge and voltage ✓: when the charge is halved, voltage is also halved so the energy is more than halved. So the time for the energy to halve is shorter. ✓

(e) In between the capacitor plates $E = \dfrac{V}{d} = \dfrac{1.2 \times 10^3}{0.35 \times 10^{-3}}$

$\quad\quad = 3.42 \times 10^6 \, N \, C^{-1}.$

$F = EQ$ ✓ $= 3.42 \times 10^6 \times 1.6 \times 10^{-19} = 5.49 \times 10^{-13} \, N$

$a = \dfrac{F}{m}$ ✓ $= \dfrac{5.49 \times 10^{-13}}{9.11 \times 10^{-31}}$ ✓ $= 6.02 \times 10^{17} \, m \, s^{-2}.$

(f) (i) $\quad v^2 = u^2 + 2as$ ✓

$\quad\quad v^2 = 0 + 2 \times 6 \times 10^{17} \times \dfrac{0.35 \times 10^{-3}}{2} = 2.1 \times 10^{14} \, ms^{-1}$

$\quad\quad v = \sqrt{2.1 \times 10^{14}} = 1.449 \times 10^7$ ✓

(ii) $\quad \frac{1}{2}mv^2 = \frac{1}{2} \times 9.11 \times 10^{-31} \times (2.1 \times 10^{14})$ ✓ $= 9.56 \times 10^{-17} \, J = 0.6 \, keV?$

An eV is the amount of energy to accelerate an electron through I V. $\dfrac{1.2 \, kV}{2} \rightarrow 0.6 \, keV$ ✓

(iii) $\quad v = u + at$ ✓

$\quad\quad 1.449 \times 10^7 = 0 + 6 \times 10^{17} \, t$ ✓

$\quad\quad t = \dfrac{1.449 \times 10^7}{6 \times 10^{17}} = 2.41 \times 10^{-11} \, s$ ✓

Examiner commentary

(a) Seren is living slightly dangerously – the first mark is often for the correct substitution in the equation – as she obtained the correct answer, this was awarded 'by implication'.

(b) (i) Writing 10–6 instead of 10^{-6} was considered a slip and ignored.

(ii) Seren has made a mistake in the power of 10. It should be 10^{-3}.

(c) Almost an ideal answer: writing $R = 670$ instead of $670 \times 10^3 \, \Omega$ on the first line was a little alarming, but Seren used the correct value later.

(d) Good reasoning – even nicer would have been to say that the time for energy to halve is half the time for the charge to halve, but this was not needed for full credit.

(e) This question was answered correctly by only a few candidates. Seren's solution is well expressed: in a 'show that' question, all the steps must be shown and fully described. It also helps to give a more precise answer than the one asked for.

(f) (i) Seren has actually made three slips of expression here – but none was considered serious enough to cause the loss of a mark:

- writing the unit of v^2 as $m s^{-1}$
- writing $m \, s^{-1}$ as ms^{-1} [i.e. milliseconds^{-1}]
- omitting the unit of v in the final answer.

(ii) Seren's lost mark is in the conversion of J to eV: She wrote '$9.56 \times 10^{-17} \, J = 0.6 \, keV$'. This is not clear enough for a 'show that' answer. We needed to see something like:

$$9.56 \times 10^{-17} \, J = \frac{9.56 \times 10^{-17} J}{1.6 \times 10^{-19} C} = 597 \, eV$$

$$= 0.6 \, keV$$

(iii) Good.

Seren scores 18 out of 20 marks.

Quickfire answers

1.5 Circular motion

① $\pi, \frac{\pi}{2}, \frac{\pi}{4}, \frac{\pi}{6},$

② 5.5π

③ (a) 50 Hz

  (b) $100\pi = 314$ s^{-1}

  (c) 75.4 m s^{-1}

④ 2.7 m s^{-2}

⑤ 97.2 kN – the rails

⑥ (a) 1.13 N

  (b) 0.30 m

  (c) 1.30 m s^{-1}

1.6 Vibrations

① 36 [s^{-2}]

② $[\omega^2] = \dfrac{[a]}{[x]} = \dfrac{\text{m s}^{-2}}{\text{m}} = \text{s}^{-2}$

  $\therefore [\omega] = \text{s}^{-1}$ QED
  Angular speed

③ 2.4 N m^{-1}

④ 0.994 m

⑤ −0.032 m

⑥ 0

⑦ $\frac{2}{3}$ s, $\frac{4}{3}$ s

⑧ (a) 0.262 m s^{-1}

  (b) −0.131 m s^{-1}

⑨ $E_P = 68$ μJ
  $E_k = 129$ μJ

⑩ (a) $\frac{1}{2}$ (b) $\frac{1}{4}$

⑪ 2.0 Hz

⑫ $T^2 = \dfrac{4\pi^2 l}{g}, \therefore l = \dfrac{g}{4\pi^2} T^2$

  $\therefore s + \dfrac{d}{2} = \dfrac{g}{4\pi^2} T^2$

  $\therefore s = \dfrac{g}{4\pi^2} T^2 - \dfrac{d}{2}$ QED

⑬ Increase the mass / choose a spring with a lower k / use 2 or 3 springs in series.

1.7 Kinetic theory

① 5.31×10^{-26} kg

② 0.0664 kg

③ 670 kPa

④ 600 K

⑤ $n = 0.024$ mol
  $N = 1.45 \times 10^{22}$

⑥ 326 kPa

⑦ 316 m s^{-1}

⑧ (a) 0.195 kg

  (b) 7.81 kg m^{-3}

  (c) 480 m s^{-1}

  (d) 296 K

  (e) 6.13×10^{-21} J

⑨ 468 m s^{-1}

1.8 Thermal physics

① (a) Heat in a sealed rigid container.

  (b) 75 J

② Zero ($pV = 400$ J at all points)

③ (a) 10.5 J

  (b) 430 J

④ (a) $\Delta T = 60$ K

  (b) $\Delta U = 37.5$ J

  (c) $W = 25$ J
    (i.e. 25 J work by the gas)

  (d) $Q = 52.5$ J
    (i.e. 52.5 J of heat into the gas)

⑤ (a) $T_B = 120$ K

  (b) $\Delta U = -360$ J

  (c) $W = 360$ J

⑥ ~350 J

⑦ Calculate W using the area under the graph, then apply $\Delta U = Q - W$ with $\Delta U = 0$.

⑧ Over AB, heat taken in has to supply work as well as internal energy. No work for BC.

⑨ 76.6 °C

⑩ To give time for equilibrium to be achieved.

⑪ $a = \dfrac{pA}{nR}$

⑫ Large extrapolation required; all the data points a long way from −273 °C.

⑬ 2.88 Ω

⑭ ~ 4 minutes

⑮ Advantage: Less percentage uncertainty in ΔT
  Disadvantage: More heat loss

⑯ Gradient $= \dfrac{VI}{cm}$

2.4 Capacitance

① (i) 2.5×10^{-9} F = 2.5 nF

  (ii) 85 V

② 4.7×10^{-11} C = 47 pC

③ 1.8 μm

④ (i) $Q = 4.47$ mC

  (ii) $U = 21.3$ mJ

  (iii) $I_0 = 40$ A

⑤ 0.55 mm

⑥ $C = 3.09$ μF; $A = 190$ m^2; not realistically.

⑦ $C = 1.0$ nF

⑧ $C = 57$ pF

⑨ (a) 7.5 mA

  (b) 15 nA

⑩ 18 μC × 37% = 6.7 μC
  → $RC = 0.7$ s (from the graph)
  → $R = 40$ kΩ (to 1 sf)

⑪ (i) Charge – switch up
      Discharge – switch down

  (ii) 0.15 s

  (iii) 2.04 mC

  (iv) 0.10 s

  (v) 25%

⑫ Because the product RC (i.e. the time constant) is small.

⑬ (a) 42 μs

  (b) 6.0 V

  (c) 8.4 V

  (d) 49 μs

⑭ $I = I_0 e^{-\frac{t}{RC}}$

⑮ (a) 7.03 V

  (b) 10.8 s

  (c) 7.2 mF

⑯ $I = I_0 e^{-\frac{t}{RC}}$; when the switch is opened.

⑰ (a) $I_0 = \dfrac{\text{cell EMF}}{R}$

(b) $V_0 = $ cell EMF

⑱ Short it out using a wire (preferably incorporating a resistor)

⑲ 5.6 s

2.6 Electrostatic and gravitational fields of force

① (a) 240 nN ↑
(b) 20 m s^{-2} ↑

② F m^{-1} = C V^{-1} m^{-1}
= C (J C^{-1})$^{-1}$ m^{-1}
= C (N m C^{-1})$^{-1}$ m^{-1}
= C^2 m^{-2} N^{-1} QED

③ × $\frac{1}{4}$

④ 445 V m^{-1}

⑤ (a) B is 6 kV higher potential
(b) − 5.4 μJ

⑥ 1.1×10^5 m s^{-1}

⑦ Gradient = − 1.0 kV m^{-1} in agreement with 1.0 kN C^{-1}.

⑧ 2.5 kV m^{-1} away from midpoint of AB

⑨ 1.5 kV m^{-1} in the direction of the vector A→B

⑩ 206 V

⑪ 0 (zero)

⑫ (a) 2.3×10^{-22} N
(b) 5.5×10^{-65} N

⑬ 5.97×10^{24} kg

⑭ 5500 kg m^{-3}

⑮ (a) − 0.63 mN kg^{-1}
(b) − 500 MJ kg^{-1}

⑯ Internal energy (of the body and atmosphere)

⑰ 2.37 km s^{-1}

⑱ 12.0

⑲ $V_{3.14}$ = − 1.34 MJ kg^{-1}
$V_{3.64}$ = − 1.37 MJ kg^{-1}

2.8 Orbits and the wider universe

① 27 years

② 6.02×10^{24} kg

③ 2.01×10^{30} kg

④ Because, as r increases, the mass (M) within r initially increases more quickly than r. Hence $v = \sqrt{\dfrac{GM}{r}}$ increases. For larger values of r, M continues to increase (it cannot decrease) but more slowly than r which limits means that v decreases.

⑤ The value of rv^2 is roughly constant, e.g.
$0.5 \times 72^2 = 2592$; 1.0×51^2
= 2601

⑥ 12.3. There is ~12× as much material as we can detect by conventional means, suggesting more than 10× as much dark matter as baryonic (normal) matter.

⑦ $\dfrac{d_{SS}}{d_{SE}} = \sqrt[3]{2} \sim 1.26$

⑧ −33.7 km s^{-1}

⑨ v_S = 471 m s^{-1};
r_S = 21 400 km

⑩ (a) 4.25×10^9 m
(b) 6.40×10^{10} m
(c) 5.98×10^{10} m
(d) 5.2×10^{28} kg
(e) 25.7 km s^{-1}

⑪ Viewing direction in the plane of the orbit (i.e. it's seen edge-on) and a circular orbit.

⑫ 1.2×10^{23} m (~ 13 million light years)

⑬ $[G] = $ N m^2 s^{-2} ≡ kg^{-1} m^3 s^{-2}

$\rho_c = \dfrac{3 \times (2.20 \times 10^{-18} \text{ s}^{-1})^2}{8\pi \times 6.67 \times 10^{-11} \text{ kg}^{-1} \text{ m}^3 \text{ s}^{-2}}$
$= 8.66 \times 10^{-27}$ kg m^{-3}

⑭ ~ 5 hydrogen atoms per m^3.

3.6 Nuclear decay

① (a) $^{241}_{95}\text{Am} \rightarrow ^{237}_{93}\text{Np} + ^4_2\text{He}$
(b) $^7_4\text{B} \rightarrow ^7_3\text{Li} + ^0_1\beta$
(c) $^{99}_{43}\text{Tc*} \rightarrow ^{99}_{43}\text{Tc} + ^0_0\gamma$

② No evidence for α – counts with and without paper are within uncertainty
Large drop (30%) in count with 2 mm Al – suggesting some β radiation
Radiation getting though Al is mainly γ – some gets through 15 cm Pb

③ The γ is very weakly ionising so will not be absorbed by the body cells.
The β radiation is more ionising and is absorbed within the body cells and can cause DNA mutations.

④ 2.1 mJ (assuming half-life ≫ 24 hours)

⑤ Background ~0.5 cps, so the 11 cps is mainly γ radiation

⑥ Not significantly – expected background ~ 600 counts in 20 min.

⑦ 0.21 decays per second

⑧ (i) $\lambda = 4.91 \times 10^{-18}$ s^{-1}
(ii) $N_0 = 6.37 \times 10^{25}$
(iii) $A_0 = 313$ MBq
(iv) $A = 39.2$ MBq
(v) $A = 144$ MBq
(vi) $t = 7.77 \times 10^9$ years

⑨ Run the experiment several times (say 3) and add the corresponding number of dice after each throw.

⑩

Throw	Number of dice			
	Run 1	Run 2	Run 3	Total
0	400	400	400	1200
1				

⑪ Fractional uncertainty

$$= \frac{\sqrt{N}}{N} = \frac{1}{\sqrt{N}}$$

⑫

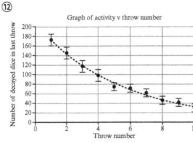

Graph of activity v throw number

(Graph: y-axis "Number of decayed dice in last throw" from 0 to 200; x-axis "Throw number" from 0 to 10)

⑬ Rather better than expected – a 'lucky' set.

⑭ Gradient = (−) 0.19 (approx); A bit larger than 0.167

⑮ $R \,/\, \text{cps} = 2.2$; $\dfrac{1}{\sqrt{R \,/\, \text{cps}}} = 0.68$

⑯ Gradient $= \dfrac{1}{\sqrt{k}}$,

so $k = \dfrac{1}{\text{gradient}^2}$

Intercept on $\dfrac{1}{\sqrt{R}}$ axis $= \dfrac{\varepsilon}{\sqrt{k}}$,

so $\varepsilon = \dfrac{\text{intercept}}{\text{gradient}}$

3.8 Nuclear energy

① 505 MeV
② (a) 1.99×10^{-26} kg
   (b) 242 u
③ (a) 59.9 MJ
   (b) 7.35 MeV = 1.18×10^{-12} J
④ 21.4 MeV = 3.43 pJ
⑤ 7.68 MeV / nucleon
⑥ 8.79 MeV / nucleon
⑦ (a) $\Delta E_p = mg\Delta h = 5250$ J

$$\therefore \Delta m = \frac{5250 \text{ J}}{(3 \times 10^8 \text{ m s}^{-1})^2}$$
$$= 5.8 \times 10^{-14} \text{ kg}$$

   (b) Because the energy to raise the rock comes from within the Earth–rock system
⑧ Because the daughter nucleus has a greater binding energy

than the sum of the binding energies of the parent nuclei.

⑨ 173 MeV
⑩ BE / nuc of U235 ~ 7.6 MeV
   BE / nuc of daughters ~ 8.4 MeV
   ∴ Energy release ~
   $(8.4 - 7.6) \times 235 = 188$ MeV

3.9 Magnetic fields

① (i) →
   (ii) →
② 2.68 mN
③ 44.0°
④ $I = nAve$

$$F = Bqv \sin\theta \times nAl$$
$$= B(nAvq)l \sin\theta$$
$$= BIl \sin\theta$$

⑤ 7.8×10^{-15} N
⑥ $Bqv = mr\omega^2$ and $v = r\omega$
   $\therefore Bqr\omega = mr\omega^2$
   $\therefore Bq = m\omega$, i.e. $\omega = \dfrac{Bq}{m}$ QED
⑦

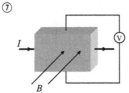

⑧ 1.2 mV m^{-1}
⑨ (a)

(diagram with voltmeter V, B field, I current)

   (b) By the FLHR, the force on the electrons is to the left, so the right-hand face has an electron deficit.
   (c) (i) 0.435 mV m^{-1}
       (ii) 1.28 mm s^{-1}
   (d) 2.1×10^{24} m^{-3}

⑩

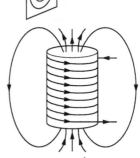

⑪ 5800 turn m^{-1}
⑫ Closer than 4.3 cm
⑬ Field due to bottom wire at the top wire = out of the diagram (grip rule)
   ∴ By FLHR force on top wire is upwards, i.e. repulsive
⑭ 1.1 mN m^{-1}
⑮ Fields on middle wire due to the other two are opposite.

$$B \propto \frac{I}{r} \text{ but } \frac{2.1}{2.5} = \frac{6.3}{7.5}, \text{ so}$$

   they cancel, i.e. resultant field $= 0$
   ∴ Force is zero.
⑯ Substitute for F and E in

$$F = Eq \rightarrow ma = \frac{V}{d}q$$
$$\therefore a = \frac{Vq}{md} \text{ QED}$$

⑰ $\frac{1}{2}m_e v^2 = eV,$

$$\rightarrow v = \sqrt{\frac{2 \times 1.6 \times 10^{-19} \times 100}{9.11 \times 10^{-31}}}$$
$$= 5.9 \times 10^6 \text{ m s}^{-1}$$

⑱ (i) 5.78 eV,
   (ii) 9.25×10^{-19} J
⑲ (i) 5.7 kV
   (ii) 9.1×10^{-16} J
⑳ It emerges from the 6th tube with 750 keV assuming the first acceleration is in the space leading up to the first tube.
㉑ (i) No change
   (ii) 3 tubes

㉒ Because they are neutral!

㉓ 3.21 GHz

㉔ Negative (the current must be clockwise around the circuit for the FLHR to predict an inward force).

㉕ (i) Doubled
   (ii) Doubled
   (iii) Quadrupled

㉖ The force on the wire is downwards (FLHR) so the force on the magnet is upwards.

㉗ (a) 0.93 (g A^{-1})
   (b) 0.15 T

㉘ B against I. The gradient should be $\dfrac{\mu_0}{2\pi a}$, but will be out by a small amount because a is not known exactly.

3.10 Electromagnetic induction

① 1.4 mWb

② 92 mWb turn

③ 15 ms

④ 5.3 cm^2

⑤ 0.11 V

⑥ The flux linkage changes giving rise to an induced emf. The direction of the change reverses, reversing the emf.

⑦ 0.126 A

⑧ 72 μA

⑨ $N\Phi$ is decreasing, so the induced current must oppose the decrease, so the flux it causes must be in the same direction, i.e. to the right. Hence the current direction must be anti-clockwise viewed from the right.

⑩ Electron flow is to the right, so the force on them is upwards giving a clockwise conventional current.

⑪ The flux linkage changes in the opposite direction.

Option A Alternating currents

① $\mathcal{E}_{in} = \omega BAN$

② 0 (zero)

③ 4.0 V

④ 32.5 V

⑤ (i) 2540 W
   (ii) 223 V

⑥ 2.0 kW

⑦ (i) 0.52 V
   (ii) 0.37 V
   (iii) 0.45 s
   (iv) 2.2 Hz

⑧

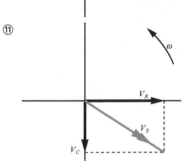

Note that the horizontal position of the trace is arbitrary

⑨ Top trace: 175 μV
   Bottom trace: −125 μV

⑩

⑪

⑫ (i) 11.9 kΩ
   (ii) 663 Ω
   (iii) 11.2 kΩ
   (iv) 0.304 mA
   (v) 89.8°
   (vi) 5.66 kHz

⑬ (i) 0.133 A
   (ii) 4.09 nF
   (iii) $V_L = V_C = 431$ V
   (iv) 0 (zero)

⑭ 0.49 mA.
The reactances of L and C are equal at the resonance frequency f_0. At $2f_0$, X_C is halved and X_L is doubled. At $0.5f_0$, X_C is doubled and X_L is halved, so the reactance (and hence the circuit impedance) is the same at the two frequencies.

⑮ f_0(max) = 5.3 MHz; f_0(min) = 530 kHz

⑯ $Q_{max} = 500$; $Q_{min} = 50$

⑰ f_0(max) = 5.3 MHz; f_0(min) = 530 kHz (same as QF 15 because L and C are the same)

⑱ $Q_{max} = 1000$; $Q_{min} = 10$

Option B Medical physics

① 60 keV

② 25 pm

③ The equation gives 8900 V. It needs to be higher because although 8900 eV is could raise an electron to the required energy level, this energy level will be full. So extra energy is required to ionise the atom.

④ 0.069 cm

⑤ 1.2 cm

⑥ 27 500

⑦ $H = 1.2$ mSv; $E = 0.06$ mSv

⑧ $H = 8.75$ mSv; $E = 0.44$ mSv

⑨ 3.5 days i.e. 12 extra hours

⑩ 1.67 ps

⑪ $[Z] = [c][\rho] = $ m s^{-1} kg m^{-3}
   $= $ kg m^{-2} s^{-1}

⑫ 5×10^{-5}

⑬ No. $\cos 15° = 0.966$ (only 3.4% error)

⑭ 14.8 cm

Option C The physics of sports

① 0.2 m

② Take moments about the bottom right corner:

$F_{min} \times 120$ cm $= 600$ N $\times 50$ cm

$$\therefore F_{min} = \frac{600 \text{ N} \times 50 \text{ cm}}{120 \text{ cm}}$$

$$= 250 \text{ N}$$

③

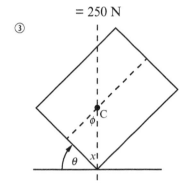

$x = 90° - \theta$ [x and θ add to a right angle]

and $x = 90° - \phi$ [angles in a right-angled triangle]

$\therefore \theta = \phi$ QED

④

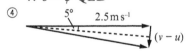

Consider the upper triangle. It is right-angled

$\therefore \frac{1}{2}|v - u| = 2.5 \sin 5°$

$\therefore |v - u| = 5.0 \sin 5°$

$= 0.44 \text{ m s}^{-1}$

Vector $v - u$ at right angles to the pecked line.

$\therefore v - u$ at 5° to the vertical.

⑤ 9.5 N

⑥ Subst into [1] →

$20 = 2v_A + 3 \times 7.2$

$$\therefore v_A = \frac{20 - 3 \times 7.2}{2}$$

$$= -0.8 \text{ m s}^{-1}$$

Subst into [2] → $8 = -v_A + 7.2$

$\therefore v_A = -8 + 7.2 = -0.8 \text{ m s}^{-1}$

⑦ Fraction lost = 21.6%

No – transferred to internal energy.

⑧ $v_A = -11.6 \text{ m s}^{-1}$;

$v_B = 4.4 \text{ m s}^{-1}$

⑨ Because the acceleration is constant (assuming negligible air resistance).

$v^2 = u^2 + 2gH$

But $u = 0$, so $v^2 = 2gH$, and hence $v = \sqrt{2gH}$ QED

⑩ So that air resistance is zero

⑪ 0.42 s

⑫ (a) 22.5 kg m^2

(b) 45 kg m^2

⑬ 1.4 kg m^2 assuming uniform sphere and cylinder

⑭ 140 kg m^2 s^{-1}

⑮ N m = kg m s^{-2} m

$= $ kg m^2 s^{-2}

Rad has no dimensions

$\therefore$ N m = kg rad^2 m^2 s^{-2}

⑯ 0.0625 [6.25%]

⑰ 30.6 m s^{-1} at 36.5° to the horizontal

[Did you remember the direction?]

⑱ The air on the top of the ball (in the diagram) stays in contact with the ball longer than on the bottom. Hence it is deflected downwards, so the ball exerts a downward force on the air (N2) and the air exerts an equal and opposite force on the ball (N3).

Option D Energy and the environment

① Area the Earth presents to the radiation $= \pi R_E^2$

$\therefore$ Power input

$= 1.37 \times 10^3 \text{ W m}^{-2} \times \pi$

$\times (6370 \times 10^3 \text{ m})^2$

$= 1.75 \times 10^{17}$ W

② Fraction $\sim 0.025\% \sim \frac{1}{4000}$

③ 1.2×10^{17} W

④ $\lambda_{max} = WT^{-1}$

$$= \frac{2.90 \times 10^{-3} \text{ m } K}{5770 \text{ K}}$$

$= 5.0 \times 10^{-7}$ m ~ 0.5 μm

⑤

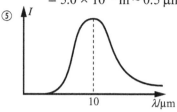

⑥ It assumes an approximately uniform temperature, i.e. the same temperature day and night.

⑦ Assuming 10% is reflected (from Fig. D1)

$\to T \sim 270$ K

⑧ 89% submerged, so very nearly!

⑨ Less radiation is reflected from the sea, so more is absorbed.

⑩ (a) 0.411 MeV (b) 6.57×10^{-14} J

⑪ 2.0×10^{10} J

[Note: only 0.5 mol of reactions]

⑫ Stefan's law → 5780 K (3 sf), so very close.

⑬ 25 kW [126 squares × 0.2 kW h square^{-1}]

⑭ (a) $v = \sqrt{2gh} = \sqrt{2 \times 9.81 \times 10}$

$= 14.0 \text{ m s}^{-1}$ (3 sf)

(b) 2.75 m^3 s^{-1}

⑮ $\frac{1}{2}mv^2 = \frac{1}{2} \times 2.75 \times 1000 \times 14^2$

$= 269\,500$ W ~ 270 kW

⑯ 46.5%

⑰ $[\tau_E] = \frac{[W]}{[P_{loss}]} = \frac{\text{J m}^{-3}}{\text{W m}^{-3}}$

$= \frac{\text{J}}{\text{J s}^{-1}} = \text{s}$

⑱ 1.2×10^8 K [120 MK]

⑲ $\frac{\Delta Q}{\Delta t} = 4 \times 10^{-4} \text{ m}^2$

$\times 200 \text{ W m}^{-1} \text{ °C}$

$\times \frac{(29.7 - 10) \text{ °C}}{0.05 \text{ m}}$

$= 31.5$ W (rounding!)

⑳ 33%

P Practical work

① (a) y against x^2
  (b) $\ln y$ against $\ln x$
  (c) $\ln y$ against x
② $106 \pm 8 \text{ cm}^3$
③ a = gradient
  b = intercept on y axis
④ $y = 6.0x^{2.5}$
⑤ $y = 22000e^{-0.15x}$
⑥ $I_{max} = 5 \text{ mA}$
  ∴ Just around the limit
⑦ The load could fall on your foot.
  The oscillating beam could strike you.
⑧ $\pm 0.368 \text{ cm}$ uncertainity
  ∴ Total length = 0.73 cm
  [or 0.7 cm, 1sf]
⑨ $\pm 0.06 \text{ s}^2$, ∴ length = 0.12 s^2

M Mathematics and data

① $[k]$ = m. We have the expression $k^2 + l^2$, so $[k]$ must have the same unit as l.
② (a) $\dfrac{4\pi^2}{g}$ (b) $\dfrac{4\pi^2 k^2}{g}$
③ $T^2 l = 218 \pm 9 \text{ s}^2 \text{ cm}$
  $l^2 = 400 \pm 8 \text{ cm}^2$
④ 0.223, 0.607, 1.65, 4.48, 12.2

⑤ $\dfrac{e^{-(x+\Delta x)}}{e^{-x}} = \dfrac{e^{-x}e^{-\Delta x}}{e^{-x}} = e^{-\Delta x}$ which
  is independent of x.
  $0.607 \ (1.65^{-1})$
⑥ $t = 0.175$ s
⑦ $1 = e^0$, ∴ $\ln 1 = \ln(e^0) = 0$
⑧ $\log_{10} 10^x = x$
  $\log_{10}ab = \log_{10}a + \log_{10}b$
  $\log_{10}\dfrac{1}{a} = -\log_{10}a$
  $\log_{10}\dfrac{a}{b} = \log_{10}a - \log_{10}b$
  $\log_{10}a^n = n\log_{10}a$
  $10^{\log_{10}x} = x$
⑨ 21 s (2 sf)
⑩ cm^{-1}
⑪ 13.8 decay times
⑫ 10 cm
⑬ $y = 12.2x^2$
⑭ A graph of $\ln\left(\dfrac{L}{L_0}\right)$ against
  $\ln\left(\dfrac{M}{M_0}\right)$ should be a straight
  line of gradient 3.5 and
  intercept $\ln 1.5$ (= 0.41) on
  the $\ln\left(\dfrac{L}{L_0}\right)$ axis.

⑮ Graph of $\ln A$ against t.
  $\lambda = -$ gradient
⑯ 0.878, 0.770, 0.479, 0.230
  0.878, 0.770, −0.479, 0.230
⑰ (a)

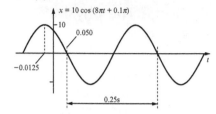

(b)

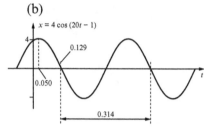

⑱ $t = 0.214$ s or 0.842 s
⑲ For the − sign:
  $v = -50\sin(10 \times 0.436 + 1)$
    $= 39.9 \text{ cm s}^{-1}$
  For the + sign:
  $v = -50\sin(10 \times 0.621 + 1)$
    $= -39.9 \text{ cm s}^{-1}$
  So the minus sign leads to the positive velocity.

Extra questions answers

1.5 Circular motion

1. (a) 3π (9.42) rad s^{-1}
   (b) 76.3 cm s^{-1}
   (c) 23.6 rad s^{-1}
   (d) 7.91 m s^{-1}
2. (a) 50π (157) rad s^{-1}
   (b) 535 N
   (c) The gravitational force on the sock is less than 1 N which is less than 0.2% of the centripetal force.
   (d) Water next to a hole has insufficient force on it to keep it inside the drum.
3. (a) Gain in KE = loss in PE
   $\therefore \frac{1}{2}mv^2 = mgl$
   $\therefore v^2 = 2gl$, $\therefore v = \sqrt{2gl}$
   (b) $2mg$
   (c) $3mg$

1.6 Vibrations

1. (a) $a \propto x$ with a negative gradient.
   (b) $k = 0.8$ N m^{-1}
   (c) $T = \pi$ (3.14) s
   (d) 0.32 s
2. (a) 50 Hz
   (b)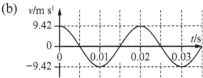
   (c) 5.54 m s^{-1}; 0.0170 s
   (d) $v = \pm 7.02$ m s^{-1}
3. (a) At the highest point, distance below attachment = 0.450 cos 15° m
   At the lowest point, distance below attachment = 0.450 m
   $\therefore$ Height loss = 0.450 − 0.450 cos 15°
   $= (1 - \cos 15°) \times 0.450$ m
   (b) $E_{k\,max} = 0.0090$ J
   (c) $A = r(\theta/\text{rad}) = 0.450 \times \dfrac{15\pi}{180} = 0.118$ m
   (d) $v_{max} = 0.550$ m s^{-1}
   (e) $E_{k\,max} = \frac{1}{2}mv_{max}^2 = \frac{1}{2} \times 0.060 \times 0.550^2$
   $= 0.0091$ J

4.

s / m	$(T/s)^2$
0.150 ± 0.001	0.635 ± 0.011
0.750 ± 0.001	3.051 ± 0.023

   (a) Max gradient = 4.097 s^2 m^{-1}
   Min gradient = 3.957 s^2 m^{-1}
   $\therefore$ gradient = 4.03 ± 0.07 s^2 m^{-1}
   (b) $g = 9.80 \pm 0.17$ m s^{-2}

1.7 Kinetic theory

1. $n = \dfrac{pV}{RT}$. So, if p, V and T are the same for two samples of a gas, the number of moles and hence the number of molecules is the same.
2. (a) For the molecule $\Delta p = -2mu$ [where p = momentum]
   $\therefore$ Change in momentum per second of gas molecules = $-2fmu$
   $\therefore$ (By N2) force exerted by the wall on the molecules = $-2fmu$
   $\therefore$ (By N3) force exerted by the molecules on the wall = $2fmu$ QED.
   (b) 1. The change in momentum (of each molecule) per collision increases.
   2. The collision frequency increases.
3. (a) 289 K
   (b) 6.0×10^{-21} J
   (c) 1340 m s^{-1}
4. (a) 1330 m s^{-1}
   (b) Collisions with other molecules transfer momentum
   (c) (i) 1.20×10^{24}
   (ii) 1300 m s^{-1}
   (iii) 1840 m s^{-1} (i.e. $\sqrt{2} \times$)

1.8 Thermal physics

1. (a) (i) Quickly compress using the piston
   (ii) Heat, e.g. using a Bunsen burner, holding piston stationary
   (b) (i) $\Delta U > 0$, ($Q = 0$), $W < 0$
   (ii) $\Delta U > 0$, $Q > 0$, ($W = 0$)

2. (a) $n = 0.100$ mol
   (b) $T = 348$ K
   (c) $\Delta U = 60$ J
   (d) Work done on gas ~ 60 J
   (e) $Q \sim 0$, so piston must have been pushed in quickly
3. (a) $T_A = 300$ K; $T_B = T_C = 1200$ K
   (b)

	AB	BC	CA	ABCA
$\Delta U / J$	450	0	-450	0
W / J	300	-555	0	-255
Q / J	750	-555	-450	-255

4. (a)

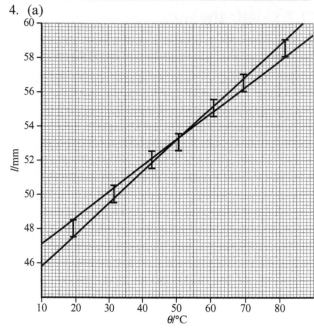

   (b) With l in mm and θ in °C:
   Steepest graph: $l = 0.1844\theta + 43.956$
   Least steep graph: $l = 0.1525\theta + 45.575$
   (c) Best fit graph:
   $t = (0.168 \pm 0.016)\theta + (44.8 \pm 0.8)$
   (d) With steepest graph: -238 °C
   With least steep graph: -299 °C
   $\therefore$ Best estimate: (-270 ± 30) °C

   > Note: as with other graph questions, your answers could differ slightly from the author's because of having to draw lines by eye and judgements in calculating gradients.

2.4 Capacitance

1. (a) (i) 2.8 nF
       (ii) 125 nC
       (iii) 2.7 μJ
       (iv) 1.87 μC (Max pd = 660 V)
   (b) $C \propto \dfrac{1}{d}$; $V_{max} \propto d$
       $\therefore Q_{max} = CV_{max} =$ constant
2. (a) 17.5 pF
   (b) 1.12 mF
   (c) 1.68 mF
3. (a) 16.2 μC
   (b) 120 kΩ
   (c) 0.324 s
   (d) 6.4 μC
   (e) 0.61 s
4. (a) 4.50 V
   (b) 8 ms
   (c) Gradient of Q v t graph = 1.0 mA at origin
       $\rightarrow R = 4.5$ kΩ
   (d) 1.8 μF
5. (a) Time constant $= RC$. The only R is the internal resistance of the supply which is very low.
   (b) $RC = 330\ \Omega \times 1.9 \times 10^{-9}$ F $= 0.495$ μs
   (c) (i) 1.6 V
       (ii) 20 nV
       (iii) 0 ($< 10^{-99}$ V)
   (d) (i) C discharges totally f times per second, $\therefore$ $I = Q_0 f$
       (ii) 18 mA
   (e) If the discharge time falls towards 1 μs the discharge will be incomplete, $\therefore I < Q_0 f$.
   (f) The following graph assumes that the switch is in contact with the upper and lower contacts for 0.5 μs with zero travel time between.

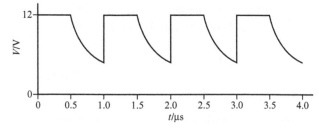

6. (a) 470 ms

   (b) $C = 2 \times$ gradient $= 56 \pm 4$ mF
   This overlaps (just) with the 47 ± 9 mF marking
   $\therefore$ Just consistent

   (c) (Using 470 ms) $10 \times$ this is 4700 ms (4.7 s)
   Fraction remaining $= 2 \times 10^{-9}$
   This is undetectably small, $\therefore$ it will have no effect on result.

2.6 Electrostatic and gravitational fields of force

1. (a) (i) 5400 V m^{-1} $\rightarrow$
       (ii) 5.14 mm s^{-2}
       (iii) 0 (zero)

   (b) (i) $x_0 = 0.23$ m; gradient of the V–x graph is zero
       (ii) $E = -\dfrac{dV}{dx}$.

       For $x < x_0$, $\dfrac{dV}{dx} < 0$; for $x > x_0$, $\dfrac{dV}{dx} > 0$

       (iii) Close to uniform because the V–x graph is a straight line
       (iv) Total charge $= (-8 + 8 - 8)$ nC
            $= -8$ nC.
       At large values of x the distance to all the charges is close to x.

   (c) Acceleration decreases to zero (at $x = x_0$) then becomes negative. Velocity increases becoming $\sim$ constant between $x = 0.2$ m and 0.3 m then decreasing gradually to zero as $x \rightarrow \infty$.

2. (a) (i) $V = -\dfrac{GM}{r}$
       (ii) $V = -\dfrac{GM}{r_s}$

   (b) For $r < r_s$, $g = -\dfrac{Gm}{r^2}$ where m is the mass within this radius.
   $m = M \times \dfrac{r^3}{a^3}$, $\therefore g = -\dfrac{GMr}{a^3} = \dfrac{r}{a}g_a$

3. (a) $v = \sqrt{\dfrac{2GM}{r}}$
   $= \sqrt{\dfrac{2 \times 6.67 \times 10^{-11} \times 1.99 \times 10^{30}}{6.96 \times 10^8}}$
   $= 618$ km s^{-1}

   (b) height above surface $= 2.32 \times 10^8$ m
   [This is 1/3 of the radius]

   (c) 2950 m ($\sim$ 3 km)

2.8 Orbits and the wider universe

1. (a) Kepler 2. In an elliptical orbit the satellite will move faster when closer to Mars, so it will not always be above the same point on the surface of Mars.

   (b) $h = 2.05 \times 10^7$ m $- 3.37 \times 10^6$ m
   $= 1.71 \times 10^7$ m

   (c) Planet must orbit above Martian equator; for any other orbital plane it will not be above a single point on the surface.

2. (a) $\dfrac{mv^2}{r} = \dfrac{GMm}{r^2}$ so $v^2 = \dfrac{GM}{r}$

   so $E_k = \frac{1}{2}mv^2 = \frac{1}{2}m\dfrac{GM}{r}$

   That is $E_k = \dfrac{GMm}{2r}$

   (b) As r decreases, E_k increases

   (c) As r decreases total (satellite) energy gets more negative, so decreases! This accords with energy being given away in collisions to particles external to the satellite.

3. (a) (i) Gravity provides the centripetal force.
   $\therefore$ For an orbiting body of mass m
   $\dfrac{mv^2}{r} = \dfrac{GMm}{r^2}$.

   Simplifying $\rightarrow v^2 = \dfrac{GM}{r}$
   $\therefore v = \sqrt{\dfrac{GM}{r}}$ QED

       (ii) $\dfrac{v_1}{v_2} = \sqrt{\dfrac{r_2}{r_1}}$

(b) (i) $\frac{v_1}{v_2} = 1.75[\pm 0.08]$; $\sqrt{\frac{r_2}{r_1}} = 1.73$

(ii) Relationship is based on a (spherically symmetric) body of fixed mass being orbited. 2 kpc is inside the 'central bulge' of the galaxy, so 'shells' of stars and other baryonic matter beyond 2 kpc from the galaxy centre don't contribute to the field at 2 kpc: the relevant baryonic mass of galaxy is much less at 2 kpc than at 6 kpc.

(iii) $8.8 [\pm 1.0] \times 10^{39}$ kg $= 4.4 [\pm 0.5] \times 10^9 \, M_\odot$

(c) v greater at all r on full line than broken, so more mass everywhere in the galaxy than the baryonic mass we know about. v increases rather than decreases with r, so far more mass in outer regions than known about.

4. (a) (i) $T = 2100$ days; $d = 4.99 \times 10^{11}$ m

(ii) $v_S = 9.15 [\pm 0.05]$ m s^{-1};
$r_S = 2.64 \times 10^8$ m;
$r \approx d = 4.99 \times 10^{11}$ m

(b) 1.19×10^{27} kg

(c) 17.3 km s^{-1}

(d) Jupiter

3.6 Nuclear decay

1. (a) E or F – must be able to penetrate packing; shouldn't need replacing often

(b) E – intensity should depend on thickness; shouldn't need replacing often

(c) B – short half-life to limit exposure; must (only) penetrate a short distance of tissue

(d) C – short half-life to limit long term danger; must penetrate significant distances of air / soil

(e) D – Long half-life to avoid need to replace; intensity affected by smoke particles; short range so doesn't affect people in the room

(f) (A or) B – need to be absorbed by gut lining to kill cells; short half-life to remove evidence.

2. Force proportional to velocity and charge.
$\therefore$ Gamma unaffected.

Radius of curvature $= \frac{mv}{Bq} = \frac{1}{Bq}\sqrt{2E_k m}$

α particles have a mass $\sim 8000\times$ that of electrons with only double the charge. So, for a given KE, the radius of curvature of α particle tracks is much greater, so the curvature is much less.

3. (a) $t_{\frac{1}{2}} = 5.27$ years

(b) $\dot{m} = 1.28 \times 10^{-7}$ kg; $N_0 = 1.28 \times 10^{18}$

(c) $A = 1.86$ GBq

(d) $A = 143$ MBq

(e) 15.8 years

(f) $t = 10.5$ years (3.3×10^8 s)

4. Age $= 12\,480$ years, $\therefore$ 10 500 BCE

5. (a) After $7\alpha + 4\beta$, $\quad Z \to Z - 7 \times 2 + 4 \times 1$
$\therefore \quad 92 \to 92 - 14 + 4 = 82$
and $\quad A \to A - 7 \times 4$
$\therefore \quad 235 \to 235 - 28 = 207$

(b) $8\alpha + 6\beta$

(c) If N atoms of U235 to start with, after 1 half-life there are $\frac{1}{2}N$ atoms of U235 and $\frac{1}{2}N$ of Pb207. $\therefore$ Ratio $= 1.00$

(d) 1400 million years

(e) 860 million years

(f) 0.14

6. (a) At larger distances the count rate goes down, so the counting time is increased to keep the total number of counts high, so that the fractional uncertainty is kept small.

(b)

d / cm	10	15	20	25	30	50	70
R / cpm	759	280	153	104	64	24	10
$1/\sqrt{R}$ / cpm	0.036	0.060	0.081	0.098	0.125	0.204	0.316

(c) Graph is a straight line – confirming the inverse-square relationship
Gradient = 0.0046.

$\therefore k = \dfrac{1}{\text{gradient}^2} = 47\,200$ cm^2 cpm

And intercept $= -0.0121$

$\therefore \varepsilon = -\dfrac{\text{intercept}}{\text{gradient}} = \dfrac{0.0121}{0.0046} = 2.6$ cm

3.8 Nuclear energy

1. (a) (i) 1.4×10^{27} u
   (ii) 2.5×10^{22} u
   (iii) 0.000 549 u
   (iv) 3.6×10^{30} u
   (v) 6.75×10^{9} u
   (b) (i) 9.3×10^{-26} kg
   (ii) 6600 kg
   (iii) 5.8×10^{-9} kg
   (iv) 1.99×10^{30} kg; the Sun
2. (a) 4.001 508 u
   (b) 6.65×10^{-27} kg
3. (a) 17.6 MeV
   (b) 0.155 MeV
   (c) 4.27 MeV
   (d) -0.09 MeV $= -90$ keV
4. 3.39×10^{14} J; 400 g of $^2_1$H required
5. (a) 6.92 MeV nuc^{-1}
   (b) 8.76 MeV nuc^{-1}
   (c) 7.52 MeV nuc^{-1}
   (d) 7.06 MeV nuc^{-1}
6. (a) The daughter nucleus has a greater BE per nucleon (and hence total binding energy) than the parent nucleus for heavy nuclei. (Same true of Be8 decay.)
   (b) Slightly less because the potential energy of $^8_4$Be is slightly greater than the potential energy of the two $^4_2$He nuclei.
   (c) Energy release is 17.7 keV but the binding energy per nucleon of $^3_1$H is greater than that of $^3_2$He.
7. ΔBE / nuc $\sim 7.6 - 7.0 = 0.6$ MeV nuc^{-1}
   There are 12 nucleons
   $\therefore$ energy release $\sim 12 \times 0.6 \sim 7$ MeV
8. (a) ΔBE/nuc ~ 0.8 MeV $\rightarrow$ energy release $= 0.8$ MeV nuc$^{-1} \times 240$ nuc $= 190$ MeV
   (b) Mass required ~ 5 kg
9. (a) β^- decay
   (b) 189 MeV
   (c) Energy in β decays $= 19.6$ MeV
   $\rightarrow$ Percentage in $\beta \sim 9.4\%$
10. 188 MeV

3.9 Magnetic fields

1. (a) Downwards
   (b) Into the diagram
   (c) Downwards
   (d) Downwards
2. (a) The direction is not constant
   (b) $Bqv = mr\omega^2$. But $v = r\omega$, $\therefore$ $Bqr\omega = mr\omega^2$
   $\therefore$ Simplifying: $\omega = \dfrac{Bq}{m}$, i.e. $f = \dfrac{Bq}{2\pi m}$
   which is independent of the speed.
   (c) Mass $= 3.01$ u, $\therefore$ either $^3_1$H$^+$ or $^3_2$He$^+$
   (d) (i) 5.0×10^7 m s^{-1}
   (ii) 39 MeV
3. (a) Bottom is +ve; FLHR
   (b) 1.03 m s^{-1}
   (c) 7.4×10^{23} m^{-3}
   (d) [p-type] semiconductor
4. (a) (i) $\Delta E_k = 15$ keV
   (ii) $\Delta v = 2.3 \times 10^5$ m s^{-1}
   (iii) 345 kHz
   (b) (i) 4.29 mT
   (ii) 39.6 keV
   (iii) 1.35×10^8 m s^{-1}
   (c) (i) Magnetic field increased
   (ii) (Assuming 4 accelerations per lap) 300 000 revolutions
   (iii) Either: Mass of a proton ~ 1 GeV
   $\therefore$ Kinetic energy $\sim 1200\times$ the mass energy
   Or: $v = \sqrt{\dfrac{2E_k}{m}}$
   $= \sqrt{\dfrac{2 \times 1.2 \times 10^{12} \times 1.6 \times 10^{-19} \text{ J}}{1.67 \times 10^{-27} \text{ kg}}}$
   $= 1.5 \times 10^{10}$ m s^{-1}
   This is greater than the speed of light.
5. (a)

$(I / \text{A})^{-1}$
0.238 ± 0.005
0.495 ± 0.010
0.719 ± 0.014
0.990 ± 0.020
1.250 ± 0.025

   (b) The author's graph gave (with l in m): $m_{max} = 26.8$; $m_{min} = 23.1$

(c) $\therefore m = 25.0 \pm 1.9$ [or 25 ± 2]
Uncertainty = 7.6% [2 sf for the moment!]
$$\frac{1}{I} = \frac{B}{F} \times l, \text{ so } m = \frac{B}{F}, \text{ leading to}$$
$B = 0.123 \pm 0.012$ T

(d) I^{-1} depends linearly on l as predicted. The whole range of the claimed flux density $(125 \pm 5$ mT) is within the experimentally determined range.

6. With B in T and x in m, the author's graph gave:
Gradient: $m_{max} = 1.21 \times 10^{-6}$;
$m_{min} = 0.83 \times 10^{-6}$
$\therefore m = (1.02 \pm 0.19) \times 10^{-6}$
Intercept: $c_{max} = -0.0008$; $c_{min} = -0.0110$
$\therefore c = 0.006 \pm 0.005$
Gradient, $m = \frac{\mu_0 I}{2\pi}$, leading to
$\mu_0 = (4.0 \pm 0.7)\pi \times 10^{-7}$ H m^{-1}
Intercept $= -d$, so $d = 6 \pm 5$ mm from the top of the phone!

3.10 Electromagnetic induction

1. (a) 46 mWb
   (b) The flux linkage is constant
   (c) 2.2×10^{-16} N, towards the viewer
   (d) Forces on electrons in the same direction on opposite sides $\therefore$ oppose each other around the rim.

2. (a) 2.8 mWb turn
   (b) 51 V
   (c) Clockwise viewed from the left

3. (a) There is a change of flux linking the circuit $(= BA)$ because the area is changing. Hence an emf is induced which causes a current to flow.
   (b) Upwards; Lenz's law
   (c) (i) 5.56 mWb
       (ii) 77 μA
   (d) The rate of change of flux is not constant but increases
   (e) (i)

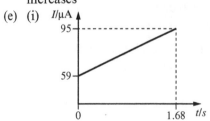

(ii)

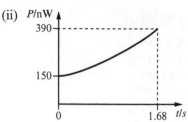

4. (a) Magnet accelerates downwards. Flux linkage increases (not linearly)
   → Increasing emf: just before magnet enters coil
   → (in coil) when magnet is central in coil, flux is constant → zero emf
   → below flux linkage decreases producing opposite emf. Magnet travelling faster so peak higher

   (b)

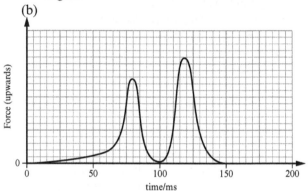

   (c) $\varepsilon_{in} = \dfrac{\Delta(BNA)}{\Delta t}$

   $\therefore B = \dfrac{\varepsilon_{in}\Delta t}{NA} = \dfrac{\text{area under graph}}{NA}$

   By square counting → 3.3 mT

5. (a) Clockwise
   (b) 15 mA

Option A Alternating currents

1. (a) 250 turns
   (b)

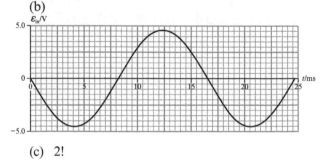

   (c) 2!

2. (a) 233 V
   (b) 15.8 A
   (c) 2610 W
   (d) 5230 W
   (e) 0 (zero)
3. (a) (i) 11.5 V, (ii) 8.13 V, (iii) 10 kHz
   (b) 8.67 mT
   (c) (i) 10 V / div
       (ii) 40 μs / div
   (d)

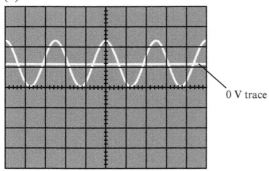

0 V trace

4. (a) (i) 83.1 Hz, (ii) 86.2 mA, (iii) 0.70,
       (iv) 3.51 V, (v) 3.51 V
   (b) (i) 60 Ω, (ii) 83.5 mA, (iii) 4.84 V,
       (iv) 4.08 V,
       (v) 2.84 V,
       (vi) The voltage leads by 14.4°
   (c) (i) – (iii) as part (b), (iv) + (v) swapped
       (vi) The current leads by 14.4°
5. (a) (i) $I_{max} = 1.88$ A (rms);
           $I_{min} = 0.375$ A (rms)
       (ii) $f_{0\,max} = 4.91$ kHz; $f_{0\,min} = 1.55$ kHz
       (iii) $Q_{max} = 1.35$; $Q_{min} = 0.085$
   (b) $X_C = 8.16$ Ω and $X_L = 14.29$ Ω, leading to:
       (i) $Z = 10.1$ Ω
       (ii) $I_{rms} = 1.49$ A
       (iii) $V_R = 11.9$ V
       (iv) $V_L = 21.3$ V
       (v) $V_C = 12.1$ V
       (vi) The voltage leads by 37.5°
   (c) Either the resistor or the capacitor will be
       shorted out by the two earth points.

Option B Medical physics

1. (a) 120 kW
   (b) 0.042%
   (c) 100 keV; 1.6×10^{-14} J
   (d) 12.4 pm
   (e) $>3.1 \times 10^{15}$ photons s^{-1}

2. (a) $\dfrac{I(\text{tissue} + \text{bone})}{I(\text{tissue})} = 9.2 \times 10^{-8}$
   (b) (i) 0.019
       (ii) 1.0 mW m^{-2}
3. (a) Efficiency $= 1.4 \times 10^{-9}$ ($1.4 \times 10^{-7}\%$)
   (b) (i) 0.060 cm^{-1}
       (ii) 9.0 W m^{-2}
       (iii) 0.38 Gy min^{-1}
       (iv) Radiation weighting factor = 1,
            ∴ Equivalent dose = 0.38 Sv min^{-1}
       (v) Assuming a tissue weighting factor of 0.12
           Time = 66 min
4. (a) Table showing acoustic impedances:

Property	Nose	Air	Gel	Skin
Z / kg m^{-2} s^{-1}	1.692×10^6	425	1.531×10^6	1.575×10^6

   From these the reflection factors for nose/air
   and air/skin are both 0.999, which means that
   without gel only 10^{-6} of the original sound
   penetrates the body.
   With gel instead of air, the reflections are
   0.0025 and 0.0002 respectively which means
   that 99.7% penetrates the body and 99.7%
   of the reflected energy from within the body
   penetrates back out to the ultrasound probe.
   (b) $\Delta f = 4300$ Hz, assuming direction of flow
   of blood is same as (or opposite to) the
   ultrasound direction.
5. (a) $f_{8.0\,T} = 340.8$ MHz; $f_{7.5\,T} = 319.5$ MHz
   (b) Advantages: 3D imaging; high definition;
       differentiates soft tissue types
       Disadvantages: High cost; allergic reactions
       to contrast agent; long duration of scan;
       discomfort / claustrophobia
   (c) Advantages: No ionising radiation / no
       carcinogenic effects; high definition
       Disadvantages: Not cancer specific (PET
       shows up cancers)
6. (a) PET/CT/X-ray – no: ionising radiation
       US – yes: no skull calcification
       MRI – yes but needs general anaesthetic.
   (b) MRI – yes (best): good contrast (but slow
       and claustrophobic)
       PET – yes: good contrast but ionising
       radiation

CT – possible: OK contrast but high ionising radiation dose

X-ray – no: poor soft tissue contrast, high ionising radiation dose

US – no: skull needs to be cut away first

(c) PET – yes: good contrast but ionising rad, slow, expensive

MRI – yes: good contrast, no ionising rad, but slow and expensive

CT – best: good contrast, quick but moderate rad dose

US – no: high reflection at tissue/air boundary

X-ray – yes: but ionising radiation and poor soft tissue contrast

(d) PET – yes: good contrast, but ionising rad, slow, expensive

MRI – yes: good contrast, no ionising rad but slow and expensive

CT – yes: good contrast, quick but moderate ionisation dosage

US – best: fluid areas will not have tissue/air boundary, so will stand out

X-ray – yes: but ionising rad and poor soft tissue contrast

(e) X-ray – yes: quickest and cheapest

PET – OK: good contrast but ionising radiation, slow and expensive

MRI – possible: slow and expensive

CT – yes: good contrast, quick but moderate ionising radiation dosage

US – possible: but poor resolution and more time consuming

Option C The physics of sports

1. (a) The angular momentum ($I\omega$) is constant. The 'tucked' moment of inertia is less, so the angular speed is greater.
    (b) Increases by a factor of 2.7
2. (a) 17.37 m s^{-1}
        NB. 4 figures not really justified.
    (b) 97%
    (c) 0.69 m s^{-2}
    (d) 1.05 rad s^{-2}
3. (a) $1.2 \times 10^{-4} \text{ kg m}^2$ (1200 g cm^2)
    (b) 0.0126 N m
    (c) 1.67 J

4. (a) 7.75 m s^{-1}
    (b) 38 W
5. So the total angular momentum is zero – to avoid applying an opposite torque to the drone.
6. (a) 16.5 m (3 sf)
    (b) Yes, by 2.2m (to 18.7 m)
    (c) Although the shot spends a shorter time in the air, the horizontal component of velocity is greater and this has a bigger effect.

Option D Energy and the environment

1. $207 \text{ K} = -66 \,°\text{C}$
2. $1.4 \times 10^{-5} \text{ m} = 14 \,\mu\text{m}$
3. (a) 39.4 J m^{-3}
    (b) $31\,500 \text{ m}^3 \text{ s}^{-1}$
    (c) 496 kW
    (d) 1.23 m
        NB. $\rho r^2 = \text{constant}$
4. Methane is a greenhouse gas, so there is a positive feedback on global warming. Similarly, Arctic ice melt increases the fraction of solar radiation absorbed.
5. (a) Speed of flow = 48.52 m s^{-1}
        Power transfer = 44.86 MW
    (b) To extract power, the water has to do work on a turbine, reducing the flow rate and hence the power transfer.
    (c) (i) 13.82 MW
        (ii) 75%
6. (a)

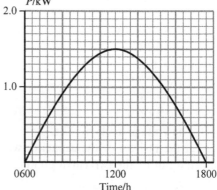

    (b) (Author's estimate) 11.4 kW h
    (c) Power in sunlight less near sunrise and sunset because of a longer travel distance through the atmosphere.

7. (a) Rate of heat loss = 628 kW
Rate of temperature decrease = $0.29\,°C\,s^{-1}$

(b) 314 W and $0.14\,°C\,s^{-1}$. They are halved because rate of heat loss is proportional to the temperature difference with the surrounding water.

(c) Question asked for a sketch graph. This is a detailed one:

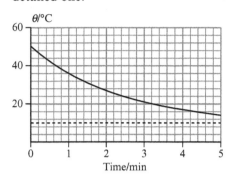

It is an exponential decay down to 10 °C because the rate of heat loss and therefore the rate of temperature drop is proportional to the temperature above 10 °C. [Half-life of $\Delta\theta$ ~1.6 min]

P Practical work

1. $28 \pm 2\,N\,m^{-1}$

2. (a) Mass makes the least contribution as $p_m = \dfrac{0.01}{95.57}$ which has the least value;

thickness is greatest as $p_t = \dfrac{0.1}{3.05}$ is bigger than any of the others.

(b) $p_l = 0.22\%$; $p_w = 0.37\%$; $p_t = 3.2(8)\%$; $p_m = 0.01\%$

(c) $\rho = 2.5 \pm 0.1\,g\,cm^{-3}$

3. Series: $R_{max} = 30.3\,\Omega$, $R_{min} = 29.7\,\Omega$,
∴ $R = 30.0 \pm 0.3\,\Omega$
Parallel: $R_{max} = 3.367\,\Omega$, $R_{min} = 3.300\,\Omega$,
∴ $R = 3.33 \pm 0.03\,\Omega$
[Do you spot the easy way of doing this?]

4. (a)

(b) A graph of (V/V) against (I/A) has a gradient −0.45 and an intercept of 3.16 on the V axis
∴ The internal resistance is 0.45 Ω and the emf is 3.16 V

(c) Eliminating I from $V = E - Ir$ and $I = \dfrac{E}{R = r}$ leads to $\dfrac{1}{V} = \dfrac{1}{E} + \dfrac{r}{ER}$

(d) A graph of $\dfrac{1}{V}$ against $\dfrac{1}{R}$ has is a straight line of gradient 0.143 and intercept 0.317 on the $\dfrac{1}{V}$ axis. ∴ $E = \dfrac{1}{0.317} = 3.15$ V and $0.317r = 0.143$ ∴ $r = 0.45\,\Omega$

(e) The scatter in the results was much smaller than the 1% expected – the results were too good!

5. (a) 0.866 ± 0.010 s

(b) 0.864 ± 0.010 s

(c) Student B's method is quicker and (if no mistakes are made, e.g. counting) just as good. Starting and stopping errors are less. Student A's method allows an estimate of uncertainty from the spread of results – which B's doesn't.

6. The author's graph of $\ln(y/m)$ against $\ln(l/m)$ has a gradient of 3.00 and intercept −4.36 on the ln y axis. Hence $n = 3.00$ and $\ln(A/m^{-2}) = -4.34$, so $A = 0.0130\,m^{-2}$

7. (a) Taking logs: $\ln y = \ln\left(\dfrac{mg}{4Eab^3}\right) + 3\ln l$

Hence the gradient is correct and the intercept is $\ln\left(\dfrac{mg}{4Eab^3}\right)$.

(b) (i) Plot a graph of y against l^3, which should be a straight line (through the origin). [Note: Other graphs are possible, e.g. $\sqrt[3]{y}$ against $l^{\frac{3}{3}}$.]

Gradient $= \dfrac{mg}{4Eab^3}$, so measure the gradient and apply:

∴ $E = \dfrac{mg}{4ab^3 \times \text{gradient}}$

(ii) [See author's graph on next page.]
(iii) Gradient $= 0.0131 \pm 0.0003$ (m^{-2})
(iv) $(1.47 \pm 0.05) \times 10^{10}$ Pa

(v) The percentage uncertainty in the gradient (2.3%) makes the biggest contribution to the overall uncertainty (3.4%) so the measurement of the depression is the chief source of uncertainty.

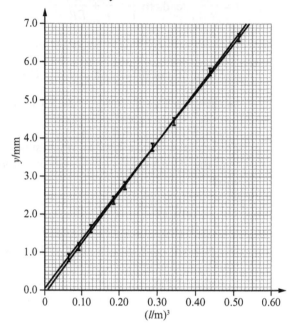

M Mathematics and data

1. T^2 against a^3, T against $a^{\frac{3}{2}}$, $T^{\frac{2}{3}}$ against a....

2. (a) Gradient $= \frac{3}{2}$; intercept $= -16.3$
   (b) Gradient $= 6.82 \times 10^{-15}$; intercept $= 0$
   (c) Gradient $= 30.2$; intercept $= 0$
   This is tricky, so a worked answer is given.
   $$T = \frac{2\pi}{\sqrt{GM}} r^{\frac{3}{2}}$$
   Putting in the values of π, G and M gives:
   $(T/\text{s}) = 8.2577 \times 10^{-8}(r/\text{m})^{\frac{3}{2}}$
   Converting to days: 1 day $= 86\,400$ s
   $$(T/\text{day}) = \frac{8.2577 \times 10^{-8}}{86\,400}(r/\text{m})^{\frac{3}{2}}$$
   $$= 9.5575 \times 10^{-13}(r/\text{m})^{\frac{3}{2}}$$
   Converting to r in 10^6 km (i.e. 10^9 m) gives
   $(T/\text{day}) = 9.5575 \times 10^{-13}(10^9\,r/10^6\,\text{km})^{\frac{3}{2}}$
   $= 9.5575 \times 10^{-13} \times (10^9)^{1.5}(r/10^6\,\text{km})^{\frac{3}{2}}$
   $= 30.2(r/10^6\,\text{km})^{\frac{3}{2}}$
   Hence the gradient is 30.2.

3. (a) $A = 5.0$ cm; $f = 3.0$ Hz; $T = 0.33$ s ; $\omega = 6\pi$ rad s$^{-1} = 18.8$ rad s^{-1}
   (b) 2.70 cm [remember radian mode]
   (c) $x = -3.17$ cm; $v = 72.9$ cm s^{-1}; $a = 11.3$ m s^{-2}
   (d) $v = \pm 56.5$ cm s^{-1}

4. (a) $\lambda = 0.0524$ h$^{-1} = 1.456 \times 10^{-5}$ s^{-1}
   (b) 8.77 GBq
   (c) $A = 1.32$ MBq; $N = 9.0 \times 10^{10}$

5. (a) $(x/\text{cm}) = 10\cos(8\pi(t/s))$ $[x = 10\cos8\pi t]$
   (b) $(v/\text{cm s}^{-1}) = -80\pi \sin(8\pi(t/s))$ $[v = -80\pi \sin8\pi t]$
   (c) $E_{k\,\text{max}} = 0.63$ J
   (d)
   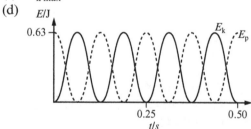

6. 12 days: **A** – one sixteenth; **B** – one quarter

Practice questions answers

① A body undergoes simple harmonic motion if its acceleration is always directed towards a fixed point and is proportional to its distance from that point. The period is the time for 1 cycle of the motion. The amplitude of the motion is the maximum distance of the body from the central point.

Commentary

A better definition of the period would be the shortest interval between times in which the body is in the same position and moving with the same velocity [i.e. same speed in the same direction].

②

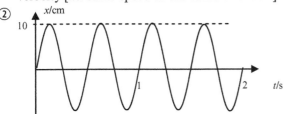

Commentary

A sketch-graph is not necessarily accurately plotted but it often needs numerical information. The axes need to be labelled and given units – but not given an accurate scale. In this case, the amplitude is 10 cm so this needs to be labelled on the x axis. The period is 0.5 s, so that information needs including on the t axis. The sine curve need not be accurately drawn but does need to be recognisable – it helps to mark the places where it crosses the axis [0, 0.25, 0.5 . . . . s].

③ A system which can oscillate undergoes free oscillations if it is displaced and then released. The frequency of the free oscillations is called the system's natural frequency. Forced oscillations occur if the system is subject to a periodic driving force – in this case the system oscillates with the frequency of the driving force.

④

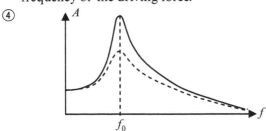

Commentary

The only feature of interest is the frequency of the peak, which is the resonance frequency. The value of this is actually slightly less than the natural frequency but this distinction can be ignored as it is so small.

The second curve should have the same amplitude at very low frequencies, the peak should be in the same place [or very slightly lower frequency], be below the first curve at all frequencies and have the same shape.

⑤ The vector sum of the momenta of the bodies of a system stays constant, providing that no resultant external force acts on the system.

Alternative answer

In any interaction between bodies, the total momentum of the bodies remains constant provided no resultant external force acts on the system.

Commentary

The use of 'vector sum' rather than just 'sum' is usually ignored: momentum is a vector quantity so can only be summed in a vector fashion. Similarly 'total momentum' is taken to imply a [vector] sum of the momenta. Writing 'provided no external forces' rather than 'no resultant external force' is not quite as good but is usually accepted.

⑥ This is the number of particles per mole of substance, which is the number of atoms in exactly 12 g of carbon-12, approximately 6.0×10^{23}.

Commentary

Going on to say what is meant by a mole is a good idea, although the question does not specifically require it.

⑦ ΔU – this is the change in internal energy of the system.

Q – this is the heat flow into the system [from the surroundings]

W – this is the work done by the system [on the surroundings]

Commentary

'Change' is by definition positive if there is an increase. Generally the change in a quantity is 'final value – initial value', i.e. $\Delta U = U_2 - U_1$. Candidates often make the mistake of writing 'heat flow into or out of the system'. This is incorrect. Writing the equation in this way requires Q to be algebraically the flow <u>into</u> the system, e.g. if 10 kJ of heat flowed out of the system, then Q would be –10 kJ, not +10 kJ. Similarly, if a gas is compressed, it does a negative amount of work <u>on</u> the surroundings.

⑧ The temperature rise, ΔT when a substance, of mass m is heated by Q is given by $Q = mc\Delta T$. The quantity c is the specific heat capacity.

The statement says that it takes 4200 J of heat per kg of water to raise its temperature by 1 K.

Commentary
An equation is a good way of defining a quantity – if all the terms in the equation are defined.

⑨ Kepler's 3rd law states that the square of the period of the orbit is proportional to the cube of the orbital radius.

Consider a planet of mass m orbiting a star of mass M, with $M \gg m$, at a distance r. The gravitational force, F, on the planet is given by: $F = \dfrac{GMm}{r^2}$.

F provides the centripetal force,

$$mr\omega^2 = mr\left(\frac{2\pi}{T}\right)^2 \text{ ie } \frac{GMm}{r^2} = mr\left(\frac{2\pi}{T}\right)^2$$

Dividing by m and re-arranging:

$$T^2 = r^3 \times \frac{4\pi^2}{GM},$$

i.e. $T^2 \propto r^3$ as required.

Commentary
Strictly, Kepler's 3rd Law refers to the semi-major axis, rather than the radius, but the above answer would be accepted.

⑩

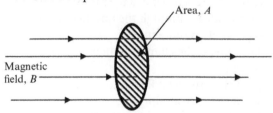

Area, A

Magnetic field, B

The magnetic flux, Φ, linking the circuit is defined by $\Phi = BA$.

⑪ A is the **activity** of a radioactive material, which is the number of radioactive decays per unit time. Its unit is the becquerel – Bq.

N is the number of undecayed nuclei in the radioactive material. It is a number and has no units.

The symbol λ is the **decay constant** of the material. Its unit is s^{-1}.

Commentary
The unit s^{-1} would be acceptable instead of Bq as the unit of activity.

Questions to test understanding

⑫ $p = mv$, so the unit of p is $\text{kg} \times \text{m s}^{-1}$, i.e. kg m s^{-1}.

$p = \dfrac{h}{\lambda}$: The unit of h is $\text{J s} = \text{kg m}^2 \text{ s}^{-2} \times \text{s}$
$$= \text{kg m}^2 \text{ s}^{-1}$$

The unit of λ is m

So the unit of $p = \dfrac{\text{kg m}^2 \text{ s}^{-1}}{\text{m}} = \text{kg m s}^{-1}$, which is the same.

Commentary
With 'show that' questions, you should always make it clear what your working means and fill in all the steps. This answer has used the fact that the joule is equivalent to $\text{kg m}^2 \text{ s}^{-2}$. If you are not confident of this, work it out using:

$$W = Fd \text{ and } F = ma.$$

⑬ (a) The speed is the same so the kinetic energy of the planet is constant. Because energy is conserved, this means that the planet's gravitational potential energy is constant and so it must always be the same distance from the star.

(b) (i) Angular speed $= \dfrac{\text{speed along orbit}}{\text{radius}}$

$$= \frac{5 \times 10^4 \text{ m s}^{-1}}{8 \times 10^{10} \text{ m}} = 6.3 \times 10^{-7} \text{ rad s}^{-1}$$

(ii) Orbital period, $T = \dfrac{2\pi}{\omega} = \dfrac{2\pi}{6.3 \times 10^{-7}}$
$$= 1 \times 10^7 \text{ s}.$$

(iii) Orbital frequency $= \dfrac{1}{T} = 1 \times 10^{-7} \text{ Hz}$

(iv) Centripetal acceleration
$$= \frac{v^2}{r} = \frac{(5 \times 10^4)^2}{8 \times 10^{10}} = 0.031 \text{ m s}^{-2}$$

(c) Centripetal acceleration
= gravitational field strength

So $0.031 \text{ m s}^{-2} = \dfrac{GM}{r^2}$

So $M = \dfrac{0.031 \times (8 \times 10^{10})^2}{6.67 \times 10^{-11}}$

$= 3.0 \times 10^{30} \text{ kg}$

⑭ The equation of the graph is
$F = -kx$, where $k = 5 \text{ N cm}^{-1} = 500 \text{ N m}^{-1}$.

So the acceleration $a = \dfrac{F}{m} = -250\,x$. Compare
this with $a = -\omega^2 x$, which is the equation for
simple harmonic motion. So the subsequent
motion is simple harmonic motion with
amplitude 10.0 cm and $\omega = \sqrt{250} = 15.8 \text{ s}^{-1}$.
The equation of shm with initial velocity zero is:
$x = A\cos(\omega t)$, so in this case $x = 10.0\cos(15.8t)$,
where x is in cm and t in seconds.
When $t = 1.5 \text{ s}$, $x = 10.0 \times \cos 23.7 = 1.38 \text{ cm}$.
The velocity, v, is given by $v = -A\omega\sin(\omega t)$
$= -158\sin 23.7 = -156 \text{ cm s}^{-1} = -1.56 \text{ m s}^{-1}$.

Commentary
For all calculations involving oscillations, it
is important that the calculator mode is 'rad'
(radians) and not 'deg' (degrees).

⑮ (a) M_1 and M_2 refer to the masses of two
point objects [i.e. ones which are very small
compared to their separation]. F is the force
of attraction between the objects, r their
separation and G the universal constant of
gravitation.

(b) The centripetal force on the Moon
$= M_{\text{Moon}}r\left(\dfrac{2\pi}{T}\right)^2$

So $\dfrac{GM_{\text{Earth}}M_{\text{Moon}}}{r^2} = M_{\text{Moon}}r\left(\dfrac{2\pi}{T}\right)^2$

And, cancelling M_{Moon} and rearranging:

$M_{\text{Earth}} = \dfrac{4\pi^2 r^3}{GT^2}$

$= \dfrac{4\pi^2 \times (384\,000 \times 10^3)^3}{6.67 \times 10^{-11} \times (27.3 \times 24 \times 3600)^2}$

$= 6.02 \times 10^{24} \text{ kg}$

(c) From Newton's law of gravitation,

$g = \dfrac{GM_{\text{Earth}}}{r^2}$,

where r = radius of the Earth.

So $M_{\text{Earth}} = \dfrac{9.81 \times (6370 \times 10^3)^2}{6.67 \times 10^{-11}}$

$= 5.97 \times 10^{24} \text{ kg}$.

(d) For a spherically symmetric object, the
gravitational field outside the object is the
same as it would be if all the mass of the
object were concentrated in a point at the
centre. In both (b) and (c) the Earth can
thus be treated as a point mass. The Moon is
quite small compared to its distance from the
Earth, so it can be treated as a point mass
for the purposes of the calculation in (b).

Commentary
It is unlikely that the examiner would require
identification of the objects as point masses in
part (a). You should study the answer to part (d)
carefully.

⑯ (a) Mass of asteroid = density × volume
$= 2500 \times \frac{4}{3}\pi \times 25^3$
$= 1.64 \times 10^8 \text{ kg}$

KE of asteroid $= \frac{1}{2} \times 1.64 \times 10^8 \times 1000^2$
$= 8.2 \times 10^{13} \text{ J}$

(b) (i) Potential due to Moon
$= -\dfrac{GM}{r} = -\dfrac{6.67 \times 10^{-11} \times 7.35 \times 10^{22}}{1.74 \times 10^6}$

$= -2.82 \times 10^6 \text{ J kg}^{-1}$

(ii) Potential due to Earth
$= -\dfrac{GM}{r} = -\dfrac{6.67 \times 10^{-11} \times 5.97 \times 10^{24}}{3.84 \times 10^8}$

$= -1.04 \times 10^6 \text{ J kg}^{-1}$

(c) Total energy = initial kinetic energy
$= 8.2 \times 10^{13} \text{ J}$

Total potential energy at impact
$= -(2.82 + 1.04) \times 10^6 \times 1.64 \times 10^8 \text{ J}$
$= -6.33 \times 10^{14} \text{ J}$

Total PE at impact + KE at impact
$= 8.2 \times 10^{13} \text{ J}$

∴ KE at impact $= 8.2 \times 10^{13} + 6.33 \times 10^{14} \text{ J}$
$= 7.15 \times 10^{14} \text{ J}$

[Note this is equivalent to about 150 000
tonnes of TNT]

⑰ First calculate the resistance of the clock:

$$R = \frac{V}{I} = \frac{3.3}{1.0 \times 10^{-6}} = 3.3 \times 10^6 \ \Omega.$$

The capacitor decay equation: $Q = Q_0 e^{-t/RC}$,
Dividing by C and remembering that $V = Q/C$:
$V = V_0 e^{-t/RC}$ with $V = 1.3$ V and $V_0 = 3.3$ V

Taking logs: $\ln V = \ln V_0 - \dfrac{t}{RC}$ so

$$\ln 1.3 = \ln 3.3 - \frac{t}{3.3 \times 10^6 \times 0.2}$$

i.e. $t = 3.3 \times 10^6 \times 0.2 \times (1.194 - 0.262)$
$= 615\,000$ s $= 170$ hours (2 sf)

Commentary
An alternative method would have been to calculate the charge Q, when the voltage was 3.3 V and 1.3 V and then to use $Q = Q_0 e^{-t/RC}$ directly.

⑱ (a) Capacitance is defined by $C = \dfrac{Q}{V}$.

So $\frac{1}{2}CV^2 = \frac{1}{2}\dfrac{Q}{V}V^2 = \frac{1}{2}QV.$

V is defined by $V = \dfrac{W}{Q}$, so the volt is

equivalent to J C^{-1}; Q has units of C,
So Units of $\frac{1}{2}QV$ = C J C^{-1} = J, which is the unit of internal energy. QED.

(b) (i) $U = \frac{1}{2}CV^2 = \frac{1}{2} \times 5 \times 3^2 = = 22.5$ J.

(ii) When the second capacitor is placed across the first, charge is transferred until the pds are equal. Because the capacitances are equal, each capacitor will have half the total charge and so the pd across each will be 1.5 V.
Total internal energy $= \frac{1}{2} \times 5 \times 1.5^2 + \frac{1}{2} \times 5 \times 1.5^2 = 5.625 + 5.625 = 11.25$ J

(iii) Half of the initial energy has been lost from the capacitors. It could be lost as: heating in the connecting wires, e-m radiation emitted from a spark when the connection is made, radio waves given out from the sudden surge of current.

Data analysis questions

⑲ (a) For y between 0.6 m and 0.3 m, T is approximately constant [at 1.90 s]. For the extreme values of y the period is greater, but the difference is very small [but more than the uncertainty].

The equation agrees with this: because of the y at the bottom of the fraction, small values of y should give large values of T; because of the y^2 at the top of the fraction, large values of y should give large values of T.

(b) Squaring the equation: $T^2 = 4\pi^2 \dfrac{k^2 + y^2}{gy}$

Multiplying by y: $yT^2 = 4\pi^2 \dfrac{k^2 + y^2}{g}$

So $yT^2 = \dfrac{4\pi^2}{g} y^2 + \dfrac{4\pi^2 k^2}{g}$

Comparing this with the straight-line relationship $y = mx + c$, this shows that a graph of yT^2 against y^2 should be a straight line with gradient $\dfrac{4\pi^2}{g}$ and intercept $\dfrac{4\pi^2 k^2}{g}$ on the yT^2 axis.

(c)

$(y/m)^2$	$(y\,T^2/m\ s^2)$
0.490 ± 0.004	2.74 ± 0.15
0.360 ± 0.004	2.14 ± 0.12
0.250 ± 0.003	1.81 ± 0.11
0.160 ± 0.002	1.40 ± 0.09
0.090 ± 0.002	1.12 ± 0.07
0.040 ± 0.001	0.92 ± 0.06

(d)

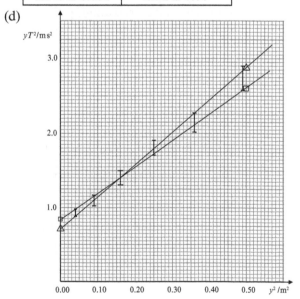

(e) Gradient of steepest line [see Δs on graph] $= \dfrac{2.87 - 0.70 \text{ m s}^2}{0.500 \text{ m}^2} = 4.34 \text{ m}^{-1}\text{ s}^2$

Gradient of least steep line

$= \dfrac{2.60 - 0.83 \text{ m s}^2}{0.500 \text{ m}^2}$

$= 3.54 \text{ m}^{-1}\text{ s}^2$

So gradient of best fit line

$= 3.94 \pm 0.40 \text{ m}^{-1}\text{ s}^2$ [i.e. 10% uncertainty]

Gradient $= \dfrac{4\pi^2}{g}$, so $g = \dfrac{4\pi^2}{\text{gradient}}$

$= 10 \pm 1 \text{ m s}^{-2}$.

Intercept $= \dfrac{0.83 + 0.70}{2} \pm \dfrac{0.83 - 0.70}{2} \text{ m}^{-1}\text{ s}^2$

$= 0.77 \pm 0.07 \text{ m}^{-1}\text{ s}^2$, i.e. 9% uncertainty.

Intercept = gradient × k^2

So $k^2 = 0.194 \pm 19\% \text{ m}^2$

So $k = 0.44 \pm 10\%$, i.e. $0.44 \pm 0.04 \text{ m}^2$.

Commentary

(a) This is a difficult pattern to interpret because of the scatter in the results. There is a small increase in T at the ends of the data. It is unlikely any examiner would be as mean as this!

(b) This is a standard piece of manipulation.

(c) This is not easy. The procedure is as follows, using the first pair of data points as an example:

$yT^2 = 0.700 \times 1.98^2 = 2.74 \text{ m s}^2$

% uncertainty in $y = \dfrac{0.003}{0.700} \times 100 = 0.43\%$

% uncertainty in $T = \dfrac{0.05}{1.98} \times 100 = 2.53\%$

So % uncertainty in $T^2 = 5.06\%$

So % uncertainty in $yT^2 = 0.43 + 5.06$

$= 5.49\%$

So uncertainty in $yT^2 = 5.49\% \times 2.74$

$= 0.15 \text{ m s}^2$

Note that it makes sense to keep several sf in the calculations of uncertainty before reducing to 1 or 2 sf at the end.

(d) The scales are chosen so that the error bars occupy at least half the available vertical height – the horizontal axis is the obvious one. The axes are labelled and the units stated.

(e) In an examination the examiner is likely to split (e) into several parts:

✓Calculation of mean gradient and uncertainty

✓Calculation of mean intercept and uncertainty

✓Calculation of g and k.

You will not be penalised for omitting units on the gradient and intercept – but you are likely to lose marks if you leave out the unit of g and k. In this case, the unit of k must be the same as the unit of y because k^2 is added to y^2.

The value of g is clearly consistent with the standard value of 9.81 m s^{-2}. The value of k for this beam should be 0.43 m^2.

⑳ (a)

1. Place the thermistor into a glass beaker containing melting ice and a thermometer, and connect it up to a resistance meter. Allow time for the temperature to become steady. Record the temperature and resistance.

2. Discard the ice and replace with water at approximately room temperature. Replace the thermistor and proceed as in step 1.

3. Heat the water using a Bunsen burner until its temperature has increased by approximately 25°C. Remove the Bunsen, allow time for equilibration and measure the resistance using the resistance meter.

4. Repeat step 3 in approximately 25°C steps up to 100°C.

Method of analysis: If $R = Ae^{\varepsilon/2kT}$, then

taking logs: $\ln R = \ln A + \dfrac{\varepsilon}{2kT}$ so a graph

of $\ln R$ against $\dfrac{1}{T}$ should be a straight line

with a gradient $\dfrac{\varepsilon}{2k}$. The value of ε can

be found from the gradient.

Results:

T/K	R/Ω	ln (R/Ω)	$(10^{-3}T/\text{K})^{-1}$
273	380	5.94	3.66
298	100	4.61	3.36
323	38.5	3.65	3.10
348	14.7	2.69	2.87
373	6.2	1.82	2.68

(b) Graph

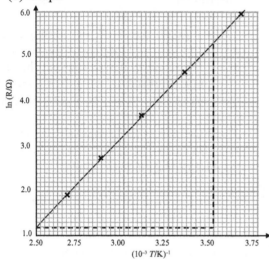

(c) The graph is a straight line, allowing for a small amount of experimental uncertainty. This supports the suggested relationship.

(d) Gradient of graph $= \dfrac{5.24 - 1.12}{(3.50 - 2.50) \times 10^{-3}\,\text{K}^{-1}}$

$= \dfrac{4.12}{1.00 \times 10^{-3}\,\text{K}^{-1}}$

$= 4.12 \times 10^3\,\text{K}$

So $\dfrac{\varepsilon}{2k} = 4.12 \times 10^3\,\text{K}$

and $\varepsilon = 2 \times 1.38 \times 10^{-23}\,\text{J K}^{-1} \times 4.12 \times 10^3\,\text{K}$

$= 1.14 \times 10^{-19}\,\text{J}$

Commentary

(a) In this case a diagram isn't necessary – all the details are clearly given in the written plan. It is not clear from the question whether the analysis of the relationship by taking logs and comparing to $y = mx + c$ is necessary here, but it needs to be done before the graph is plotted.

(b) Be careful when putting in units where logs are involved. The log of a quantity has no unit. The safest way is always to put the units with the variable – in this case ln (R/Ω). Similarly with $1/T$: write this as $1/(T/\text{K})$ or $(T/\text{K})^{-1}$. Alternatively the unit of ln R could be omitted and $\dfrac{1}{T}(\text{K}^{-1})$ written.

The graph needs careful thought – the points should occupy as much of the grid as possible. Particularly with log graphs, there is no need to include 0 on either axis unless the intercept needs to be found.

(c) No comment.

(d) The points used for calculating the gradient must be identified clearly, either by drawing the triangle as shown here, or by labelling the points – see question 19.

The unit of the gradient does not need to be included but the unit of the derived quantity, ε in this case, must be given.

Index